W9-ADZ-417

Gas Metal Arc Welding handbook

by

William H. Minnick
Associate Professor
Palomar College
San Marcos, California

South Holland, Illinois
THE GOODHEART-WILLCOX COMPANY, INC.
Publishers

About the Author

Realizing the need for a specialized type of welding text for instructors and students, Bill Minnick has drawn on his many years of experience as a welder, welding engineer, and Community College instructor to develop this text for training future welders.

The author's career in industry moved from welder to welding process engineer to welding engineer supervisor. He has had experience welding on jet engines, missiles, pressure vessels, and atomic reactors. During his industrial career, he developed the welding procedures and welded the first titanium pressure vessel for the Atlas missile program.

The author has written many technical papers for various welding publications. These articles included research and development for welding exotic materials and modification of existing welding processes for automatic and robotic applications. He is also the author of the GAS TUNGSTEN ARC WELDING HANDBOOK.

Mr. Minnick developed welding certificate and degree programs and taught all phases of welding and metallurgy in community colleges for over twenty years. His vast experiences shine through in this useful text.

The Publishers

Library of Congress Catalog Card Number 90-22906
International Standard Book Number 0-87006-867-9

1234567890-91-9876543210

Library of Congress Cataloging in Publication Data

Minnick, William H.
Gas metal arc welding handbook / by William H. Minnick.

p. cm.
Includes index.
ISBN 0-87006-867-9
1. Gas metal arc welding—Handbooks, manuals, etc. I. Title.
TK4660.M528 1991
671.5'212--dc20 90-22906
CIP

INTRODUCTION

Gas metal arc welding has long been an important welding process. Improvements in the process as well as refinements in the equipment over the years have established its position as a major industrial welding system.

GMAW HANDBOOK provides a simple but complete introduction to gas metal arc welding. It covers principles, equipment, modes of operation, and safety in a straightforward, no-nonsense manner.

A careful study of the text and practice of the welding procedures will provide you with the knowledge and skill you need to secure employment as a welder. The organization of chapters leads you step-by-step through the principles and practice of GMAW. New material is introduced as you need it. Topics covered include:

- Basic operation of each component of the system.
- Safe practices directed specifically toward working with electricity, shielding gases, and other hazards common to welding.
- Various types of welds and weld joints.
- Welding techniques and procedures for several different metals.
- Weld defects and how to avoid them.
- Special procedures and techniques for welding autobodies, trucks, and other vehicles.

For a beginning student of welding, the book provides the background needed for success in a welding career. But advanced students will value it for its detailed coverage on how to weld various metals using the different modes common to the industry. For them it will be a valuable "refresher" course.

William H. Minnick

CONTENTS

Chapter 1

GAS METAL ARC WELDING PROCESS

After reading this chapter, you will be able to:

- Define GMAW.
- List advantages and disadvantages of GMAW.
- Explain the difference between semiautomatic and automatic welding.
- Name applications where GMAW is used in industry.

DEFINITION

As defined by the American Welding Society, gas metal arc welding is an electric arc welding process. It fuses (melts) metallic parts by heating them with an arc between a continuous filler metal electrode and the work. The electrode is consumed (used up) in the process. A shielding gas protects the electrode and the molten weld metal from being contaminated by the surrounding atmosphere. Fig. 1-1 illustrates a basic gas metal arc welding operation.

PROCESS NAMES

The original development of the process led to patents by the Air Reduction Co. (AIRCO), and it was called metal inert gas welding or MIG. Many other names have been used by various manufacturers who developed variations of the process. These include:

1. Aircomatic, Air Reduction Co.
2. Sigma, Linde Co.
3. Microwire, Hobart Co.
4. Millermatic, Miller Electric Co.
5. Dipmatic, Miller Electric Co.
6. Buried Arc CO2, Miller Electric Co.

During the development years of the process, only inert gases (argon and helium) were used. The process designation, MIG (Metal Inert Gas), identified the gases as inert (chemically inactive). Other gases which were not inert came to be used in the process. The identification of the process was then changed to GMAW. This meant gases such as carbon dioxide and oxygen could be included with the GMAW definition.

ADVANTAGES AND DISADVANTAGES

GMAW has grown in use through the years since its inception. Today, because of many improvements and applications, it has become one of the major welding processes.

Advantages

1. Useful for joining major commercial metals including: carbon and alloy steel, stainless steels, aluminum, magnesium, copper, bronze, and titanium.

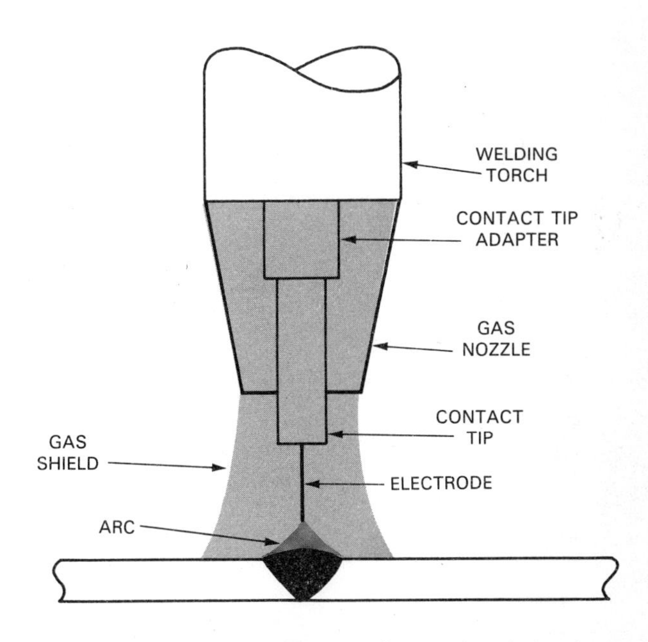

Fig. 1-1. The end of the welding torch contains the various components required to conduct electrical power to the electrode and to shield the molten metal from the atmosphere.

2. Produces high quality welds.
3. In some cases, usable in all positions.
4. Produces no slag during welding.
5. High metal deposition rate.
6. Delivers approximately 95 percent deposition efficiency.
7. In some cases, may be used to bridge gaps.
8. Welding techniques are easily learned.

Disadvantages

1. More equipment is required than for Shielded Metal Arc Welding.
2. Less portable.
3. Requires a higher initial equipment investment.
4. Subject to wind drafts which may reduce the shielding of the molten metal and cause contamination of the weld.
5. Requires various types of high cost shielding gases.
6. Welds are prone to have cold starts.
7. Cannot separate wire feed and current level.
8. Smaller diameter electrode wires are costly.

MODES OF OPERATIONS

1. When the process is done with the welder manipulating a manual torch or a portable gun, the operation is known as SEMIAUTOMATIC. A manual (semiautomatic) welding operation is shown in Fig. 1-2.
2. When the process is done by a machine or with a robot, as in some production welding, the operation is called AUTOMATIC. An automatic machine is shown in Fig. 1-3.

Fig. 1-2. This welder is using a hand torch to perform a manual (semiautomatic) welding operation. (Lincoln Electric Co.)

POWER SUPPLIES

Power supplies provide the proper type of electrical current to melt the filler material. The designation for this type of electrical current is DCRP (Direct Current Reverse Polarity) or DCEP (Direct Current Electrode Positive). A typical power supply is shown in Fig. 1-2.

WIRE FEEDER

This machine is an electric mechanical device that feeds the required amount of filler wire at a fixed rate of speed throughout the weld operation. A typical wire feeder is shown in Fig. 1-2.

Fig. 1-3. Automatic equipment reduces welding time with higher quality welds. (Miller Electric Co.)

TORCHES AND GUNS

Torches are held by the welder or a machine during the welding operation. They contain components (parts) that transfer electrical currents to the wire. Triggers or switches start and stop the operation. A gas nozzle attached to the end of the torch directs the shielding gas around the weld. This protects the molten metal from oxidation. Oxidation is a reaction caused by the introduction of oxygen to a substance. In welding, oxidation can adversely affect the strength of a weld. Fig. 1-2 shows a typical welding torch used by a welder in a weld operation.

SHIELDING GAS

A supply of the proper type of shielding gas and regulation equipment is required to admit a continuous flow of gas during the entire welding operation. Fig. 1-2 shows a supply cylinder of gas and the regulation equipment.

SYSTEMS

Components are combined into systems to operate at the proper time to make a satisfactory weld. Many different types of systems can be designed for either manual or automatic operation.

A typical semiautomatic welding system, shown in Fig. 1-4, is used for the welding of metals of all types and thicknesses.

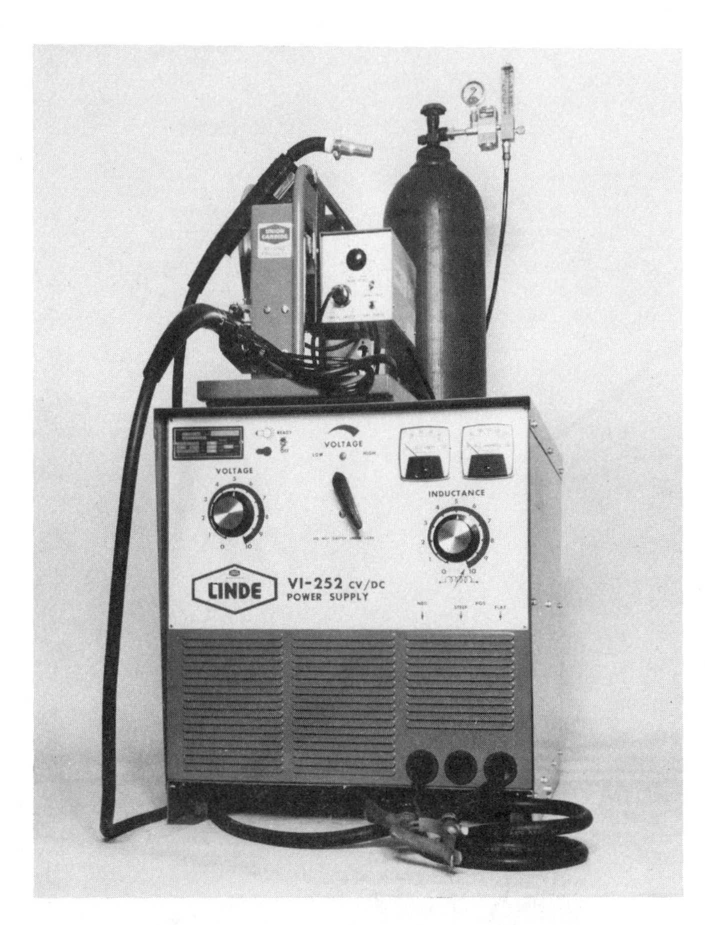

Fig. 1-4. This system can use small or large diameter wire electrodes for welding thin or heavy material with different welding modes. (Linde Co.)

A typical semiautomatic welding system, shown in Fig. 1-5, is used for the welding of thin gauge steels used in unibody automobiles.

A typical robot welding system is shown in Fig. 1-6. It can be used for welding various metals.

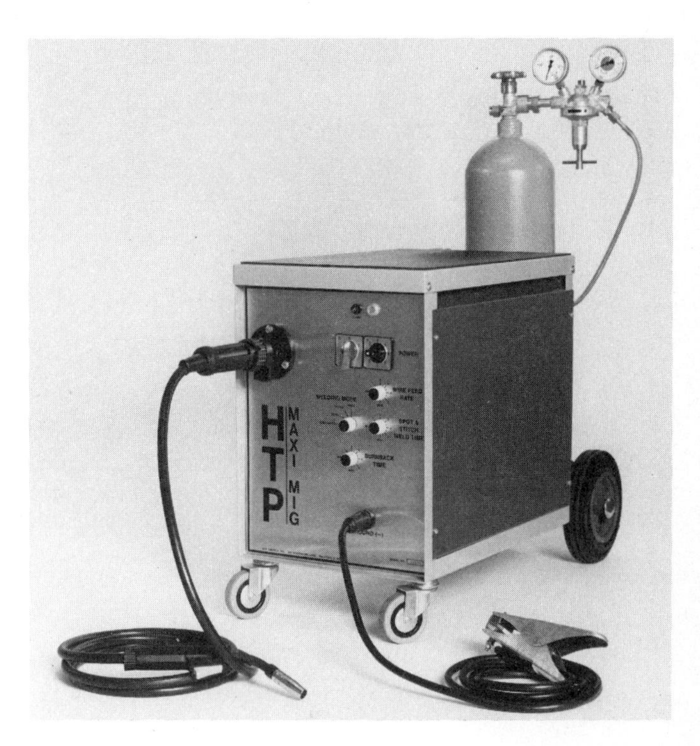

Fig. 1-5. Various controls have been added to this machine for spot, plug, and seam welding thin gauge materials. (HTP America, Inc.)

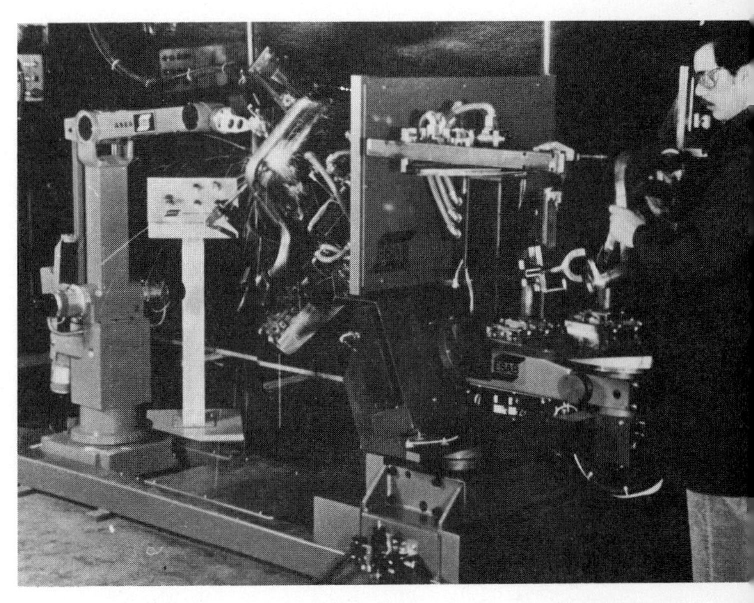

Fig. 1-6. Robots are being used to increase productivity and reduce cost. This operator is loading components for the next welding operation while another set of parts is being welded. (ESAB Welding Products, Inc.)

APPLICATIONS

There are many applications of the GMAW process in the industrial field. It may be used with various types of metals. Some of the many applications include:

1. Sheet metal assemblies.
2. Automobiles and trucks.
3. Tankers and trailers.
4. Lawn and garden equipment.
5. Motorcycles and off-road equipment.
6. Campers and recreation vehicles.
7. Structural steel assemblies.
8. Pipelines.
9. Rockets and missiles.
10. Ships, boats, and barges.
11. Pressure vessels.

REVIEW QUESTIONS—Chapter 1

1. GMAW has been defined as an electrical arc welding process by the ________ ________ ________. The process uses a ________ electrode.
2. The process was originally developed by the ________ Company and was called ________ welding.
3. The letters MIG defined the process as ________ ________ ________ welding. This was later changed to ________ ________ ________ ________ as other gases were used with the process that were not inert.
4. The two other gases that were added to the inert gases are ________ and ________ ________.
5. The manual operation of the process is often called a ________ operation.
6. GMAW always is done in the direct current electrode ________ mode, which may be termed ________, or direct current ________ ________ also referred to as ________.
7. The machine which feeds the filler wire is a electric mechanical device called a ________ ________.
8. The shielding gas which is used to prevent ________ of the molten metal is directed to the weld zone by a ________ ________.
9. The shielding gas must flow ________ during the welding operation.
10. When the welding equipment is connected together properly it is called a ________.
11. The ________ process is used in many applications on ________ types of metals.

Chapter 2

GMAW PROCESS OPERATION AND SAFETY

After studying this chapter, you will be able to:

- Name six major parameters and variables controlled by the welder.
- Discuss four basic components of a GMAW system.
- Display safe welding practices in the shop.
- Eliminate unsafe conditions in the shop.

In GMAW a consumable electrode is fed through a welding torch. Heat from an electric current melts the electrode and the base metal to make a fusion weld. The electrical current comes from the power supply. This current, also known as direct current, always moves from the power supply through the workpiece to the electrode. Electrical flow in this direction is called reverse polarity, or DCRP. (Polarity means direction in which electricity flows.) It may also be termed electrode positive or DCEP.

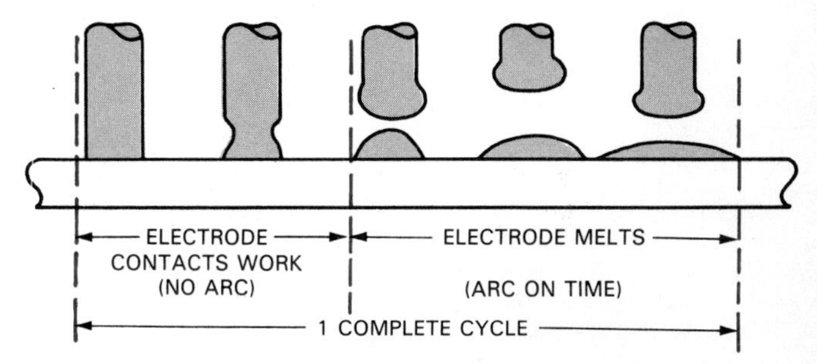

Fig. 2-1. This mode of operation is called "short-arc" as the electrode actually touches the workpiece and melts about 20 to 200 times per second.

MODES

Various modes of metal deposition (depositing) are used depending on the type of material, thickness, and type of joint to be used. The modes include:

1. Fig. 2-1 illustrates how the electrode is fed into the workpiece, short circuits, burns off, and deposits melted metal on the work. This mode of deposition is termed SHORT-ARC or SHORT CIRCUITING ARC and has many advantages and disadvantages.

   The advantages include:

   A. Low heat input.
   B. Welds in all positions.
   C. Welds thin gauge materials.
   D. Can be used to bridge gaps in the weld joint.

   The disadvantages include:

   A. The process uses small diameter wires which are expensive.
   B. Spatter may be a serious problem with various gases or techniques.
   C. Cold laps may develop in the weld due to the low heat input.

Fig. 2-2. Since the molten metal is detached from the electrode in small drops and directed onto the workpiece, this mode is termed "spray-arc."

2. Fig. 2-2 illustrates how the electrode is fed into the weld zone and is melted off above the workpiece

and is then deposited into the work. This mode of deposition is called SPRAY-ARC. It also has many advantages and disadvantages.

The advantages include:

A. Very little or no spatter.
B. High weld metal deposition.
C. Faster welding speed.

The disadvantages are:

A. The weld puddle is very large.
B. Can only be used in the flat position when welding steel groove welds.
C. Can only be used in the flat and horizontal position when welding steel fillet welds.

3. Fig. 2-3 illustrates how the electrode burns off above the workpiece in a globular pattern and is then deposited into the work. This mode of deposition is termed GLOBULAR and has many advantages and disadvantages.

The advantages include:

A. May be used in all positions.
B. High deposition rate.
C. High quality welds are possible.

The disadvantages include:

A. Considerable amount of spatter in some applications.
B. Some smoke and fumes.
C. Rough weld bead appearance.

The proper use of any metal deposition mode or variation requires that the welding equipment be set up in a special manner for each specific application. To properly set up the main characteristics for the weld, a study must be made of the required weld and the machine must be set accordingly. These characteristics are called PARAMETERS and VARIABLES in the welding trade. These will vary depending on such items as type of metal, thickness of the weld joint, weld joint design, and weld quality required

Some of the major parameters and variables are:

1. Amount of weld current.
2. Amount of arc voltage. (The gap between the electrode and workpiece.)
3. Inductance.
4. Slope.
5. Type of shielding gas, flow rate, and nozzle size.
6. Electrode diameter, wire speed, contact tip size, and electrode stickout.

The reference section of this text lists the initial parameters for various types and thicknesses of metal.

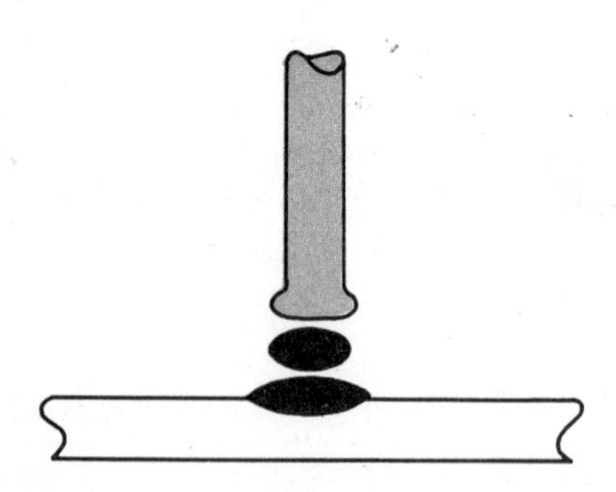

Fig. 2-3. "Globular" type deposition has a large globule of metal which is molten at the end of the electrode. Upon reaching a specific size, the globule detaches from the end of the wire and deposits on the workpiece.

POWER SUPPLIES

The basic power supply used for GMAW is designed for this type of welding process. It is called a constant potential (CP) or a constant voltage (CV) power supply. These machines supply varying amounts of amperage to maintain the preset welding voltage (arc gap). Welding voltage is set on the machine prior to starting or during the welding operation.

In some cases, a constant current (CC), which may also be termed a variable voltage (VV), machine may be used. However, these machines require additional components in the system for the process to work properly.

The term, power supplies, may in some cases include:

1. Constant voltage power supply as shown in Fig. 2-4.

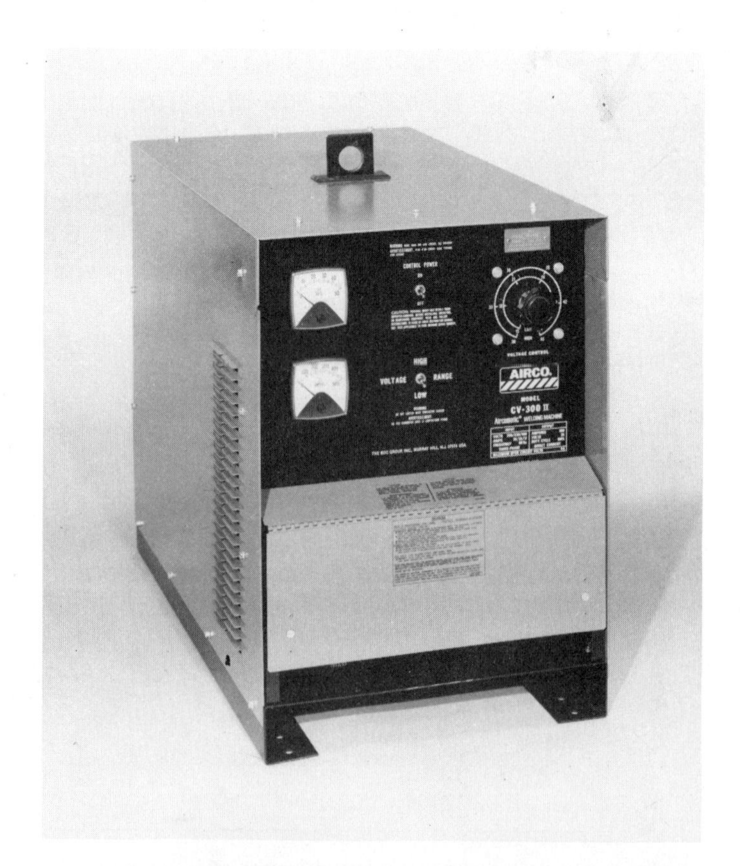

Fig. 2-4. A constant voltage power supply regulates amperage output automatically to maintain the arc voltage (gap) which is set on the power supply. (Air Reduction Co.)

2. Constant voltage power suply combined with a wire feeder in a single unit as shown in Fig. 2-5.
3. Constant current power supply coupled with a wire feeder with special circuits for wire feed is shown in Fig. 2-6.
4. Constant current (CC) and constant voltage (CV) motor generators are available for field welding where utility power is not available. A field generator is shown in Fig. 2-7. These units are either gasoline or diesel engine powered.

WIRE FEEDERS

Wire feeding equipment feeds filler wire from the supply spool, through the cable, to the torch, and finally to the arc. It may be located in different areas, or it may be located on another piece of equipment used in the process. Fig. 1-4 shows a wire feeder located separately from the power supply.

A combination unit which has a power supply and a wire feeder in a single unit is shown in Fig. 2-5.

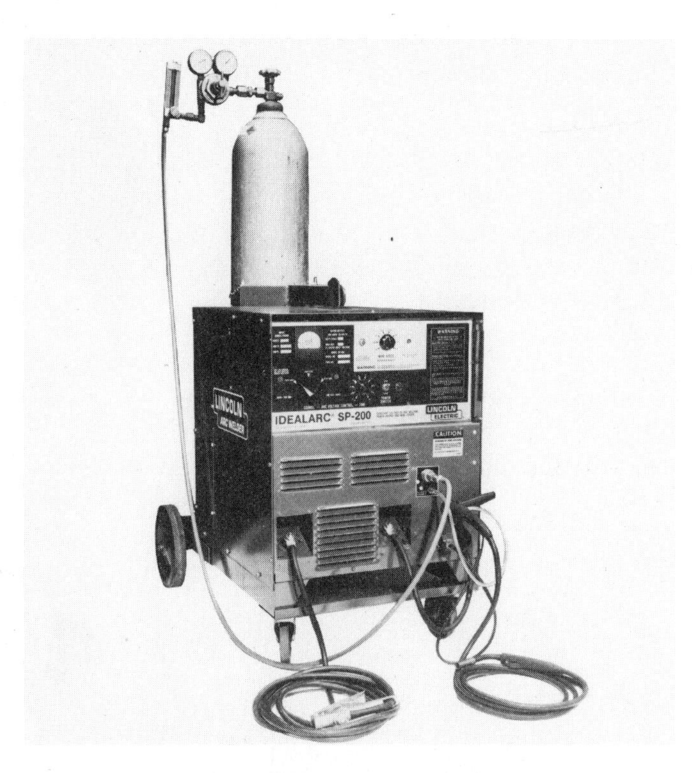

Fig. 2-5. This "welding machine" is actually a power supply and, within the machine, is a wire supply and feeder. The unit is compact and easy to move on the jobsite. (Lincoln Electric Co.)

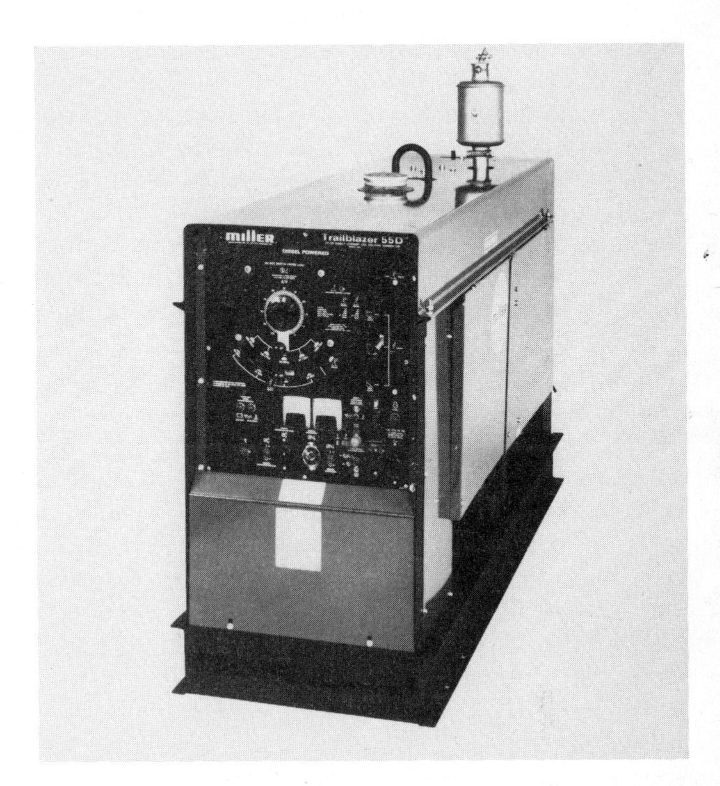

Fig. 2-7. Field units may be a single or dual process welding power source. The operator should be thoroughly familiar with this equipment before use. (Miller Electric Co.)

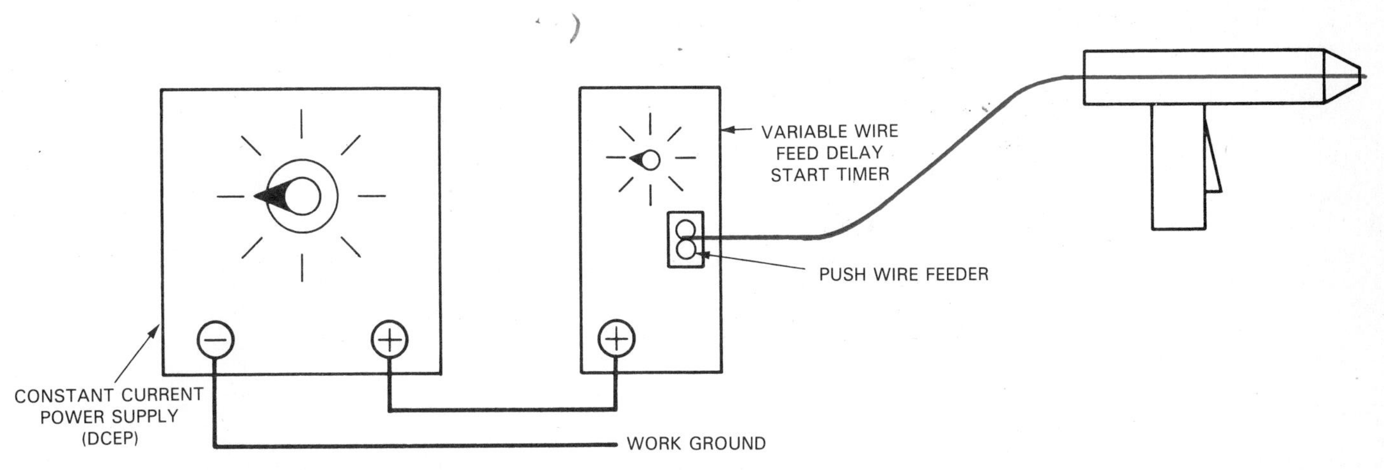

Fig. 2-6. Constant current power supplies (SMAW welding power supplies) can be used for GMAW. However, they must have a special wire feeder for delay and regulation of the wire speed at the start of the weld operation.

Shown in Fig. 2-8 is a portable gun type unit which contains both the wire feeder and the welding torch. This unit is often called a MIG gun.

GAS SUPPLY AND REGULATION

A shielding gas must be used with the welding process to avoid contamination of the welding electrode and the weld, Argon, carbon dioxide, helium, and oxygen are the most common gases used. They are supplied to the user either as a single gas or a combination of mixed gases. Depending on the amount of gas used, they may be supplied as a liquid and then converted to a gas prior to use. They may also be supplied in pressurized cylinders of various sizes.

Since the cylinders contain gases under high pressure, regulators must be used to reduce the pressure to a workable amount. A flowmeter is then used to establish the amount of gas flow to the weld area. Where a high pressure system is used, a regulator/flowmeter, as shown in Fig. 2-5, is needed. The flow of gas is controlled in cubic feet per hour (CFH) or in liters per minute (LPM).

For normal operation, sufficient gas must be delivered to the weld area for proper shielding. This will prevent contamination of the electrode and the weld metal. Too little gas flow allows the atmosphere (air) to enter the weld zone and contaminate the weld. Too much gas flow causes turbulence in the gas envelope (shield) allowing the atmosphere to enter and mix with the gas shield. (TURBULENCE means great changes in speed and direction of the flow.) Whenever atmospheric gas enter the shielding gas envelope weld quality lessens.

Electrically operated valves, commonly called solenoids, are included in the system to start and stop the flow of shielding gases from the main supply to the welding torch. These valves are actuated by closing the switch on a manual torch or closing a switch in the arc starting circuit. (These valves are usually located in the wire feeder.)

WELDING TORCHES

The GMAW process makes use of special welding torches. They are rated for current capacity that they carry, and for the length of time that they are used at that amperage. Torches used for light duty are gas cooled, and torches used for heavy duty are water cooled. The torch is constructed from materials that protect the user from electrical shock and provide a means to conduct electrical current to the welding wire as it passes through the torch.

Gas nozzles of various designs attach to the torch to direct the shielding gas around the wire and the molten metal. A typical GMAW manual torch is shown in Fig. 2-9. A typical GMAW torch, called a push-pull torch, is shown in Fig. 2-10. A portable gun called a SPOOL ON GUN is shown in Fig. 2-8. Where the welding is done with a machine or robot, the torch shown in Fig. 2-11 might be used.

Fig. 2-8. An electric motor drives the wire feed roller system at a speed set on the adjustable rheostat on the gun casing. (CK Systematics)

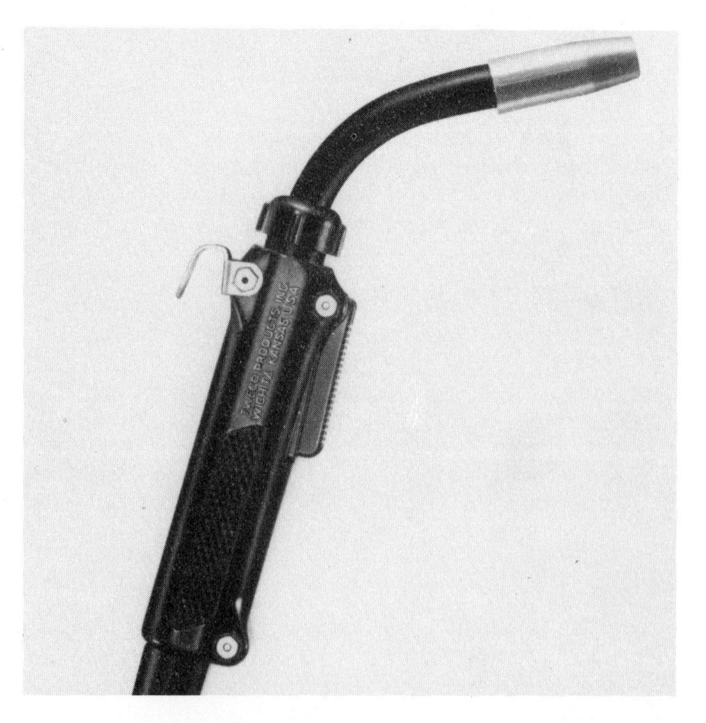

Fig. 2-9. This manual GMAW torch is gas cooled by the passage of shielding gas through the assembly. (Twecco Products, Inc.)

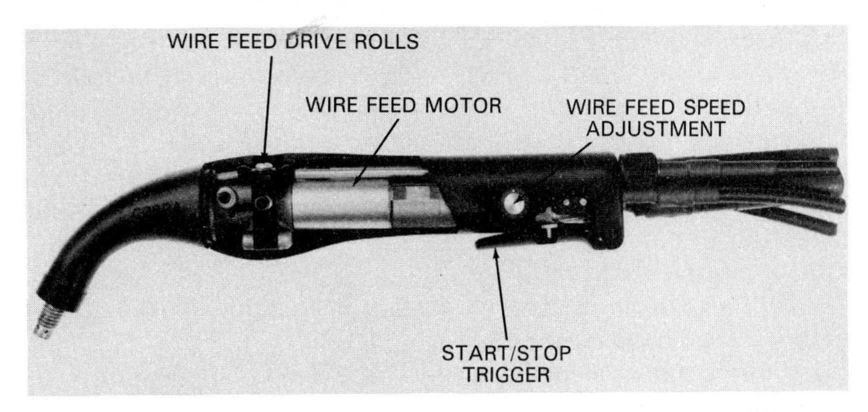

Fig. 2-10. An electrical motor and drive rolls within the torch body pull the filler material from the supply spool. Water hoses are attached to the rear of the torch body which give this torch a very high duty cycle. (M&K Products, Inc.)

Fig. 2-11. Machine torches are often water cooled to allow higher amperage duty cycles. This torch is mounted directly onto the wire feeder and a wire straightener is mounted on top of the feeder. A straightener is used to remove some of the curve of the wire prior to its entering the feed rolls. (Linde Co.)

GMAW SAFETY

GMAW is a skill which may be performed safely with minimum risk if the welder uses common sense and follows rules safely. It is recommended that you establish safety habits as you work in this industrial area. Check your equipment regularly and be sure that your environment is safe. Safety in GMAW covers four major areas and includes:

1. ELECTRICAL CURRENT. Primary current to the electrical powered welding machine is usually 208 volts AC or more. This amount of voltage can cause extreme shock to the body and possible death. For this reason, one should abide by the following rules:
   A. Never install fuses of a higher amperage rating than specified.
   B. Always ground the welding machine.
   C. Install electrical components in compliance with all codes.
   D. Be certain that all electrical connections are tight.
   E. Never open a welding machine cabinet when it is operating.
   F. Lock primary voltage switches open and remove fuses when working on electrical components inside the welding machine.
   G. Welding current supplied by the power supply has a maximum of 75 open circuit volts, however, most machines have from 30 to 50 volts. At this low voltage, the possibility of lethal shock is very small. It will still produce a good shock. To reduce the possibility of this occurrence:
      a. Keep the welding power supply dry.
      b. Keep the power cable, ground cable, and torch dry.
      c. Do not weld in a damp area. If you must work in a damp area, wear rubber boots and gloves.
      d. Make sure the ground clamp is securely attached to the power supply and the work-piece.
2. SHIELDING GASES. Some of the gases used in GMAW are produced and distributed to the user

in two forms, high pressure and liquid. All storage vessels used for these gases are approved by the Department of Transportation or the Interstate Commerce Commission and are so stamped on the vessels cylinder walls as shown in Fig. 2-12.

Some of the gases used in GMAW are inert, colorless, and tasteless. Therefore, special precautions must be taken when using them. None of the gases are toxic, however, some can cause asphyxiation (suffocation) in a confined area without adequate ventilation. Any atmosphere that does not contain at least 18 percent oxygen can cause dizziness, unconsciousness, or even death.

Shielding gases cannot be detected by the human senses and can be inhaled like air. Never enter any tank or pit where gases may be present until the area is purged (cleaned) with air and checked for oxygen content.

High pressure gas cylinders contain gases under very high pressure (approximately 2,000 to 4,000 PSI) and must be handled with extreme care. Each of the following rules should be followed:

A. Store all cylinders in the vertical position.
B. Secure all cylinders with safety chains or cables. See Fig. 2-13.
C. Do not use as rollers.
D. Know the contents before use. Fig. 2-14 shows a label indicating the type of gas in the cylinder.
E. Keep the protective cap in place until ready to use.
F. Do not move a cylinder without the protective cap in place. Always use a cylinder cart, a cylinder cap, and safety chains to move a cylinder.
G. Check the outlet threads before attaching the regulator and clean the valve opening by

Fig. 2-13. Secure all cylinders with safety chains.

Fig. 2-12. This cylinder has been made and tested to a Department of Transportation (DOT) specification. The letter "T" indicates the amount of gas which it will hold. The name imprinted is that of the owner.

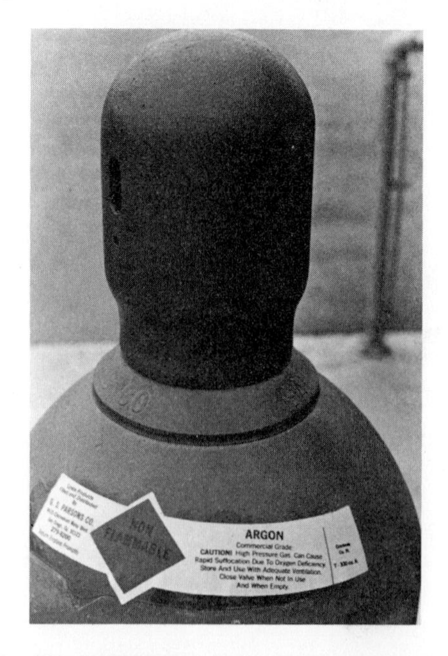

Fig. 2-14. The label indicates the type of gas stored in this cylinder. Notice the caution label and cylinder contents in cubic feet is also indicated on the label. The safety cap is in place until ready to use.

"cracking" (slightly opening and closing) the cylinder valves as shown in Fig. 2-15.

H. Use only equipment designed for high pressure usage.
I. Attach regulator securely and mount flowmeter in the vertical position as shown in Fig. 2-16.
J. Open cylinder outlet valve slowly.
K. Always stand aside when opening a cylinder tank valve.
L. When cylinder is empty, close the valve, replace the safety cap, and mark in chalk "MT."
M. Never tamper with a leaky valve, return the cylinder to the supplier.

Liquefied gas cylinders are commonly called DEWARS and are basically vacuum bottles. The gas has been reduced to a liquid at the supplier's plant for ease in handling, shipping, and storage. Conversion from a liquid to a gas is done by heat exchangers within or on the Dewars system. The following safety rules apply to a Dewars system:

A. Cylinders must be kept in an upright position.
B. Cylinders must always be moved on cylinder carts. Dewars are very heavy and difficult to handle.
C. Always use the proper equipment when installing or connecting Dewars.
D. Do not interchange equipment components.
E. Liquid gases are extremely cold and can cause severe frostbite when exposed to the eyes or the skin. Do not touch frosted pipes or valves.

3. WELDING ENVIRONMENT SAFETY RULES.
   A. Keep the welding area clean.
   B. Keep combustibles out of the welding area.
   C. Maintain good ventilation in the weld area.
   D. Repair or replace worn or frayed ground or power cables.
   E. Make sure the part to be welded is securely grounded.
   F. Welding helmets should have no light leaks.
   G. Use a proper shade number lens to protect the eyes from arc radiation. (See Reference Section for proper lens shade to use.)
   H. Wear safety glasses when grinding or power brushing.
   I. Wear tinted safety glasses when others are tack welding or welding near you.
   J. Use safety screens or shields to protect your work area.
   K. Wear proper clothing. Your entire body should be covered to protect you from arc radiation.
   L. When welding on cadmium coated steels, copper, beryllium copper, use special ventilation to remove fumes and vapors from your area.
   M. Do not weld near triachloroethylene vapor degreasers. The arc changes the vapor to phosgene gas. A sweet taste in your mouth indicates that

Fig. 2-15. "Cracking" a cylinder before attaching a regulator blows out any dirt in the valve opening.

Fig. 2-16. This regulator and flowmeter have been attached to the cylinder properly. Note the pressure relief vent on the regulator and cylinder. These devices operate to release excess pressure in the cylinder. Never tamper with these devices.

this gas is being formed.

4. SPECIAL PRECAUTION AREAS.
   A. Fires may be started by the welder in a number of ways, such as combustible materials, fuel gases, short circuits, improper ground connections, etc. Make sure that you do not start a fire. If you do, then make sure it is out before you leave the area.
   B. Never weld on a container that has held a fuel until you are sure that it has been purged with an inert gas and tested for fume content.
   C. Never enter a vessel or confined space that has purged with an inert gas until the unit or space is checked with an oxygen analyzer.
   D. Never use oxygen in place of compressed air. Oxygen supports combustion and will make a fire burn violently.
   E. Power wire brushes are very dangerous as they expel broken pieces of wire. Always wear safety glasses or safety shields when using this type of equipment.
   F. Be alert to the clamping operation when working with mechanical, hydraulic, or air clamps on tools, jigs, and fixtures. Serious injury may result if parts of the body are exposed to the clamp action.

REVIEW QUESTIONS—Chapter 2

1. The electrical flow of current for the GMAW process is always through the ________ to the electrode.
2. With the electrical flow in this direction, it is always termed ________ or ________.
3. When the welding process is set up with the electrode contacting the workpiece and melting off, this is called the ________ mode.
4. Short-arc mode is useful in welding ________ ________ materials since the arc is not on during the entire process.
5. When the electrode melts off in droplets above the workpiece, this mode is called ________.
6. Spray-arc mode has a very high metal deposition. However, it can only be used in the ________ and ________ position.
7. When the process is set up to produce a large ball on the end of the electrode before melting, it is termed ________ deposition.
8. The basic welding power supply for globular deposition is called a constant ________ or constant ________ power supply.
9. Constant current power supplies used in GMAW require a wire feeder with ________ ________ for feeding of the electrode.
10. ________ and ________ are characteristics which are used to establish machine and weld settings for a specific mode of welding.
11. Portable gun type wire feeders which contain the welding torch are often called ________ guns.
12. The most common gases used with the GMAW process are ________, ________, ________ ________, and ________.
13. Gases used from a high pressure cylinder are reduced to a lower pressure with a ________.
14. A ________ is used to control the amount of gas delivered to the weld zone through the shielding nozzle.
15. The amount of gas delivered through the gas nozzle is measured in ________ ________ ________ ________ or ________ ________ ________.
16. Too much gas flow through the nozzle causes ________ in the gas flow which may effect the weld quality.
17. Welding torches are ________ cooled for light duty and ________ cooled for heavy duty welding.

SAFETY TEST

1. When installing electrically powered welding machines, all ________ must be complied with for safety purposes.
2. Never install machines with ________ higher than those specified by the manufacturer.
3. Be certain that all connections are ________ and always ________ the welding machine.
4. Do not work on the inside of the machine until you are certain that the power cannot be turned on by someone else. ________ switches open and ________ the fuses.
5. Always keep the power supply, power cables, and torches ________.
6. When welding in a ________ area, always wear rubber boots and gloves.
7. Store all gas cylinders in the ________ position and ________ them to a solid object.
8. Always check to make sure you ________ the contents of the cylinder before use.
9. Always have the cylinder ________ in place until ready to use.
10. Always use a ________ cart to move a cylinder with the cylinder ________ installed.
11. Use only ________ designed for high pressure gases when using cylinder gases.
12. Always open cylinder outlet valve ________ and never ________ in front of the pressure gauge.
13. Never ________ with a leaky cylinder valve. Return the cylinder to the supplier.
14. Always keep ________ liquid gas cylinders in the upright position.
15. The liquid gases are extremely ________ and can cause frostbite when contact is made with frosted pipes.

16. Fires may start in the weld area when ________ are present.
17. Always make sure that the part to be welded is securely ________.
18. Always use a protective hood when welding with the proper ________ shade.
19. Always wear ________ clothing to shield your body from arc radiation and sparks from the molten pool.
20. Special ________ is required when welding on cadmium coated steels, copper, or beryllium copper.
21. Never weld on a vessel which has contained a fuel until it has been ________ clean and tested for fume content.
22. Never enter a vessel which has been purged with inert gas until you have tested the area with an ________ analyzer.
23. Never use ________ in place of compressed air for ________ reason.
24. Always wear protective ________ or ________ when working with power wire brushes.
25. Always be alert to your surroundings and to the possible dangers of injury when working on tools with ________ action.

Chapter 3

EQUIPMENT SETUP AND CONTROL

After studying this chapter, you will be able to:
- Name four power supply specifications to which the welder must conform.
- Describe the effects of adjusting open circuit voltage, arc voltage, slope, and inductance power supply controls.
- List nine typical maintenance guidelines.
- Discuss both major and auxiliary controls of wire feeders.
- Discuss maintenance of the wire feed system.
- Tell how to maintain cables, torches, and guns.

Setting up the equipment correctly for the GMAW process is essential. Proper setup not only guarantees intended performance standards, but also insures the welder that work can be done safely without equipment malfunctions.

POWER SUPPLIES

Power supplies are specially designed machines that produce welding current for melting the welding electrode at a low voltage. The equipment must be able to control the operation in the areas of:
1. Input voltage (primary voltage).
2. Open circuit voltage.
3. Output ratings and performance.
4. Duty cycle.

The National Electrical Manufacturers' Association has established specification EW-1 Electric Welding Apparatus for control of these areas.

Power Supply Specifications

Each power supply is designed for specific purposes. Therefore, limitations are established for the proper operation of the machine. The specification areas for machines using utility power include:
1. PRIMARY POWER TYPE, VOLTAGES, AND CYCLES. This includes alternating current single or three phase power at 110, 208, 230, or 460 volts and generally 60 Hertz (cycles).
2. PRIMARY POWER FUSING. The fuse sizes are specified on the machine and in the instruction manual for the individual machine. These limits should never be exceeded.
3. RATED WELDING AMPERES. The rated welding amperes are specified by the machine's manufacturer. These amounts of current should not be exceeded since the cooling system cannot carry away the excessive heat generated in the machine.
4. DUTY CYCLE. All welding power supplies are designed to operate for a specific time period at a specific load. The design considerations include:
   A. Size of internal wiring.
   B. Type of internal components.
   C. Insulation of internal components.
   D. Amount of cooling required.

   The duty cycles range from 20 to 100 percent. They may be made to a company specification or the NEMA EW-1 specification. This specification establishes that each 10 percentage points represents one minute of operation in a 10 minute period. Fig. 3-1 shows the cycle and time period for a NEMA rated machine.

For example, the duty cycle might rate a 150 ampere welder with a 30 percent duty cycle to weld at 150 amperes for three minutes. The machine then has to idle for seven minutes. This allows internal components to cool properly before resuming welding.

Machines that are not made to NEMA specifications have various duty cycles and time periods. The duty cycles and time periods should be listed on the machine or in the instruction manual. NEVER EXCEED THE EQUIPMENT MANUFACTURER'S DUTY CYCLE REQUIREMENTS.

DUTY CYCLE	NUMBER OF MINUTES MACHINE MAY BE OPERATED AT RATED LOAD IN A 10 MINUTE PERIOD
100%	FULL TIME
60%	6
50%	5
40%	4
30%	3
20%	2

Fig. 3-1. Power supply duty cycles limit the number of minutes the unit may be operated at rated load.

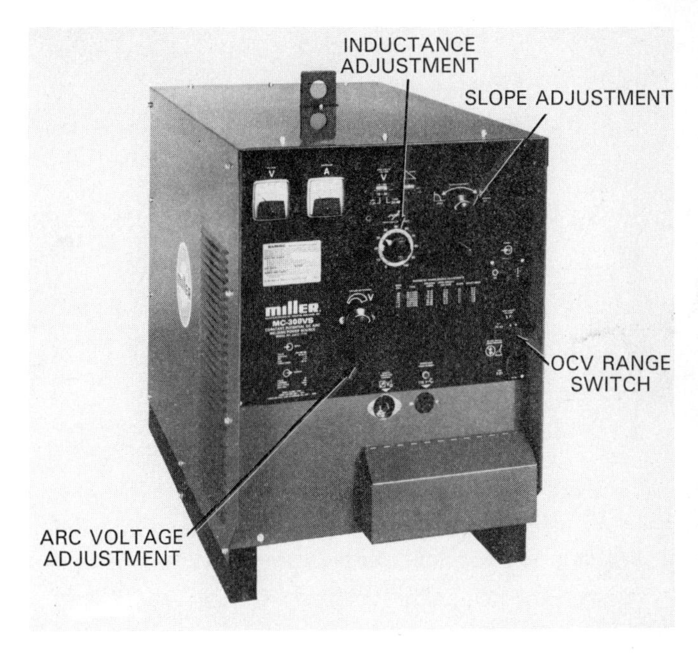

Fig. 3-2. This machine is rated at 300 amperes direct current with adjustable voltage, inductance, and slope. (Miller Electric Co.)

Power Supply Controls

Depending on the type of power supply, the type and number of controls provided for the welder may range from a tap and a simple rheostat to a multitude of controls that vary each function. A small autobody type machine may only have a tap connection for the voltage setting and a wire feed adjustment for wire speed. Another machine may be used for making very high quality welds and would need a number of controls to meet its requirements. The various controls that may be located on a machine, as shown in Fig. 3-2, include:

1. OPEN CIRCUIT VOLTAGE (OCV). Most GMAW power supplies produce a range of open circuit voltages to a high near 80 volts. This range of voltages may be controlled on the machines by changing taps, switches, or levers that are usually located on the front of the machine. As a general rule, the short-arc mode of welding requires a settime on the low end of the OCV scale and the spray-arc mode requires a setting on the middle to upper end of the scale.

   When setting up a welding power source, the welder actually sets the open circuit voltage. After the weld is started, the actual arc or load voltage is established. The actual OCV may be established on machines that are equipped with voltmeters by following these steps:

   A. Turn power supply on.
   B. Release idler roller pressure on the wire feeder to prevent wire feeding.
   C. Place voltage range switch in the desired location. (Check machine manual to select range of open circuit voltages desired.)
   D. Hold torch away from ground or workpiece and energize contactor switch on gun. (When you depress the switch, the contactor allows current to flow to the electrode tip and the wire. An arc will be made if the contact tip or wire touches the ground.)
   E. Observe the voltmeter and adjust the fine tuning voltage control to the desired open circuit voltage. THIS IS THE OPEN CIRCUIT VOLTAGE; THE ARC VOLTAGE WILL BE APPROXIMATELY 2 TO 3 VOLTS LOWER FOR EACH 100 AMPERES.
   F. Release the trigger or switch and reset tension on the wire feeder idler roller.

2. ARC VOLTAGE. Arc voltage is also called load voltage and is the actual arc gap established during the welding period. Adjust the fine voltage control (this is the same control that was used to set the open circuit voltage) to the voltage desired during welding.
3. SLOPE. The word slope in GMAW refers to the slant of the volt ampere curve and the operating characteristics of the power supply under load. In many machines, the slant of the volt ampere curve is automatically set as you change the open circuit voltage. In others, the curve can be changed by the operator for different modes of welding. The curve is known as either flat or steep mode. These curves are shown in Fig. 3-3.

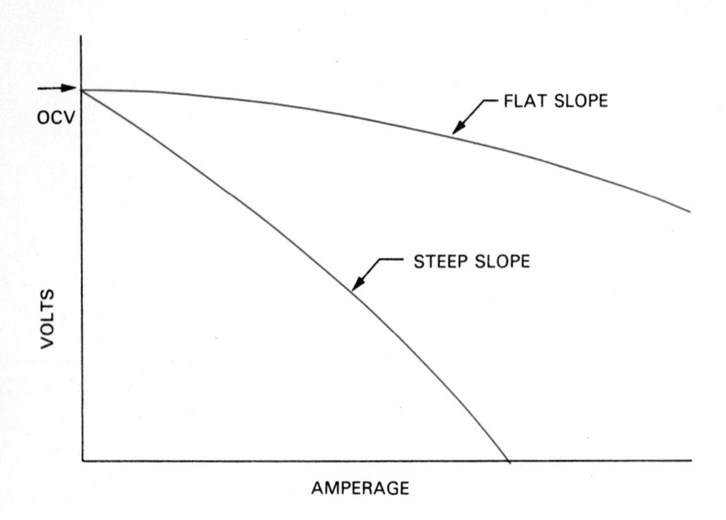

Fig. 3-3. The steep slope setting is normally used for short-arc welding. The flat slope setting is normally used for spray-arc welding.

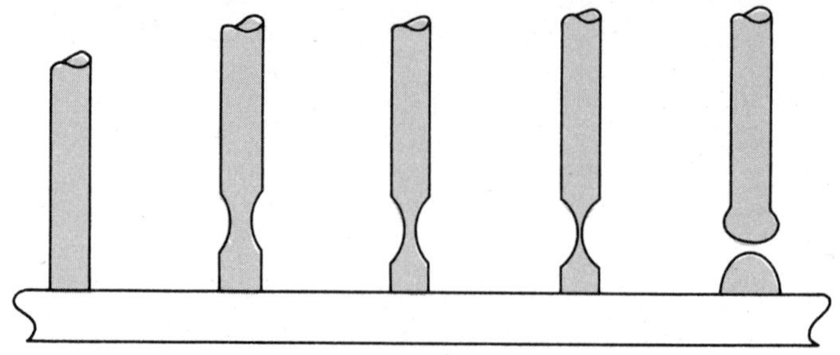

Fig. 3-4. As the electrical current flows through the electrode, heating take place until the wire melts off the end. Each cycle is completed many times per second during the welding operation. This action may be compared to a fuse in an electrical circuit. When more amperage than the wire can carry is placed in the circuit, the fuse melts.

When using a steep curve, as shown in Fig. 3-3, there is not enough current to melt the electrode at the proper time in the short-arc cycle. The arc will not start properly and the wire will stub out on the workpiece. In this case, the slope of the curve has to be decreased (flattened) to operate properly.

A machine setup with a rather flat curve as shown in Fig. 3-3 has too much current available for the short-arc cycle. As a result, the wire will be blasted off of the end, resulting in considerable spatter. In this case, the slope requires changing to a steeper slope condition.

When adjusting the slope of the volt ampere curve on the power supply, always adjust the slope so that the parting of the molten drop is smooth with a minimum of spatter.

Remember, the main function of the power supply is to provide amperage to maintain the preset arc gap or voltage as selected for welding. For this reason the machine is termed a CONSTANT VOLTAGE or a CONSTANT POTENTIAL power supply.

4. INDUCTANCE (PINCH EFFECT). When using the short-arc mode in GMAW, the separation of the molten drops of metal from the electrode is controlled by the squeezing forces exerted on the electrode due to the current flowing through it. Fig. 3-4 shows how the pinch effect operates. This is called inductance by some machine manufacturers or pinch effect by others.

   By adjusting the level of pinch effect, the machine controls the rate of current rise when the electrode contacts the workpiece and the short circuit starts. If the current flows rapidly through the electrode, the drop of metal is squeezed off rapidly and causes spatter. If inductance is added to the circuit and the current is not applied as rapidly, the number of short circuits per second will decrease, and the "arc on" time will increase. This will result in a more fluid puddle, a smoother weld crown, and less spatter.

In GMAW spray-arc mode, some inductance is beneficial in the starting of the arc process. This limits the explosive starts by slowing down the rate of current rise at the start of the cycle.

As a general rule, both the amount of short circuit current and the amount of inductance needed for the ideal pinch effect are increased as the electrode diameter is increased.

When setting up the machine for and during welding with the short-arc mode, the following guide may be useful:

1. Maximum inductance (minimum pinch).
   A. More penetration.
   B. More fluid puddle.
   C. Flatter weld.
   D. Smoother bead.
2. Minimum inductance (maximum pinch).
   A. More convex bead.
   B. Increased spatter.
   C. Colder arc.

Power Supply Installation

Power supplies should be installed in an area that is free of dust, dirt, fumes, and where the machine heat can escape. Dirt and heat will cause a power supply to overheat and ruin the internal components. The area selected for installation should remain free of objects blocking the free flow of air into and out of the machine. The machine should not be exposed to moisture, as electronic components may pick up moisture and fail. Most machine manufacturers require a defined space around the power supply for air circulation, and warranties are voided if these areas are not provided.

Utility power supplies various input voltages and 60 cycles (Hertz) as requested by the user. Machines using other than 60 cycle power are specially made at the factory for this requirement. The required fuse size for the incoming power is always shown on the data label, see Fig. 3-5. The fuse panels should always be close to the power supply. This makes is possible to disconnect the main power in an emergency. A typical power supply installation is shown in Fig. 3-6.

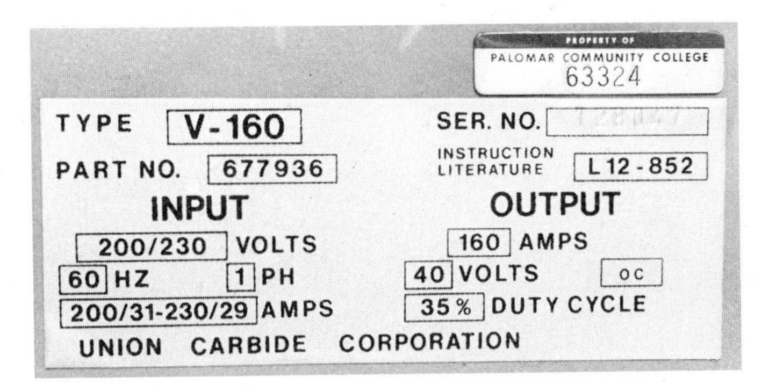

Fig. 3-5. The data label lists the requirements for incoming power and the rated output power for power supply. (Linde Co.)

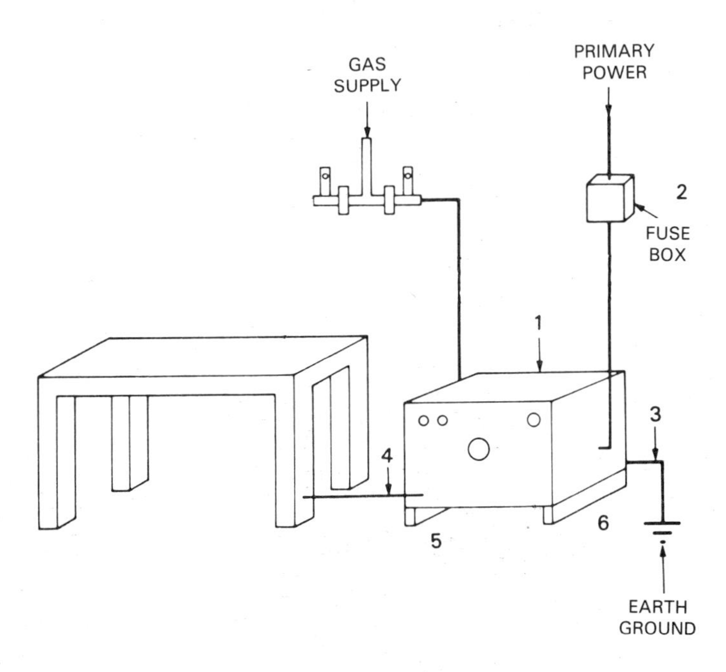

1. LOCATE POWER SUPPLY AWAY FROM WALLS FOR PROPER AIR FLOW.
2. LOCATE PRIMARY FUSE BOX NEAR POWER SUPPLY.
3. CONNECT POWER SUPPLY FRAME TO EARTH GROUND.
4. CONNECT WORK GROUND WITH 2/0 MIN-4/0 MAX CABLE.
5. KEEP POWER SUPPLY DRY.
6. KEEP AREA NEAR POWER SUPPLY CLEAN.

Fig. 3-6. Typical power supply installation includes a fuse box mounted near the welder. An earth ground from the power supply protects the welder from primary line high voltage.

Machines that do not match the utility power voltage may be used if a step-up or step-down transformer is used. Fig. 3-7 shows a step-down transformer installed for this use.

Manufacturers have excellent warranties on their products. Using a power supply voltage or fuses other than the ones stated on the data label or instruction manual may nullify the warranty.

Power Supply Maintenance

Given reasonable care and routine maintenance, a welding power supply will operate for many satisfactory hours before repairs are required. Modern methods of insulating transformers, using solid state designed components, and using good basic design have extended the hours of operation of a modern power supply. The manufacturer's instructions should always be followed for periodic inspection of the power supply. (Turn machine off and disconnect at fuse box before working inside machine.) The following points should be observed along with the manufacturer's instructions:

1. Cleaning or blowing out the unit should be done on a periodic basis. Use only dry, filtered, compressed air, nitrogen gas, or an electrical non-conducting cleaner.(Always use a face shield when using compressed air or gas.)
2. Check all terminals for loose connections.
3. Lubricate fan and motor bearings where required.
4. Check mechanical arms and switches for freedom of movement. Mechanical connections may be lightly greased if required.
5. Terminal blocks for cable connections or tap connectors should be clean and tight. If corroded, they will restrict the flow of electrical current. They may be cleaned by wire brushing or rubbing with an abrasive pad.

Fig. 3-7. These transformers change incoming primary power to match the power supply requirements.

6. Motor generator brushes should be checked and replaced when worn beyond the manufacturer's tolerance. Worn brushes wear armatures, and worn armatures must then be machined for proper operation.
7. Lubrication, inspection, and adjustment of portable gasoline and diesel power supplies requires close attention to the manufacturer's instructions. This applies to both the engine and the welder. Operation of these units usually occurs under adverse conditions. Improper maintenance will reduce the capacity and operation of the unit.
8. Be extremely careful when fueling gasoline or diesel engines.
9. Use only authorized replacement parts.

WIRE FEEDERS

Most of the feeders use 110 VAC power which is provided to the machine by a connection in the power supply. If this connection is used, the wire feeder is turned on or off when the power supply is operating or not operating.

Wire feeders used on push-pull systems or guns will use 24 volt direct current motors. These motors drive units installed in hand held torches or guns for precision drive of the wire and safety of the welder.

Types of Wire Feeders

Gas Metal Arc Welding uses three basic types of wire feeders. They include:

1. Push type. This type of feeder pushes the wire from the feeder through the cable to the torch. An example is shown in Fig. 3-8.
2. Pull type. This method usually is used when operating with a spool on a gun operation. An example is shown in Fig. 3-9.
3. Push-pull type. This type pushes the wire from the feeder through the cable to a separate wire feeder mounted in the torch handle. See Fig. 3-10. This type of system is used for welding with the soft wires as these wires will buckle in the torch liner if pushed long distances. The drive unit in the feeder is called a slave unit and only pushes the wire to the torch or gun feed rolls. The gun wire feeder then pulls

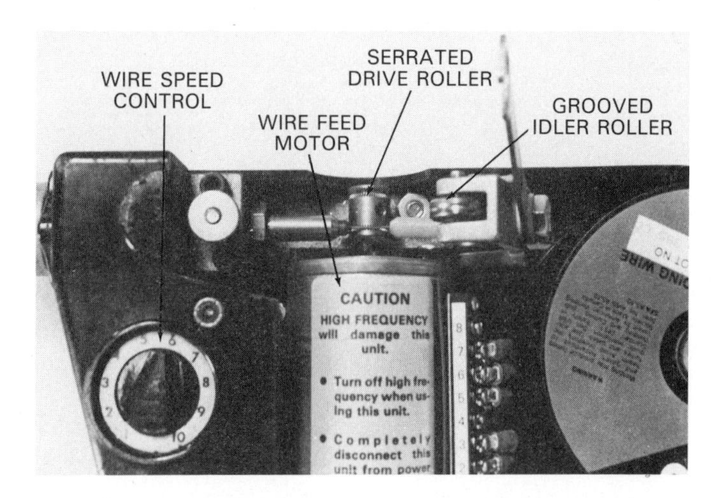

Fig. 3-9. The drive rolls are mounted in the gun head to feed the wire from the supply spool; therefore, they are called pull type feeders. The drive roller is serrated to grip the wire. (Miller Electric Co.)

Fig. 3-8. Wire feeders of this type are called push type feeders because they push the wire to the welding torch. Note the safety instructions on the unit. (Airco)

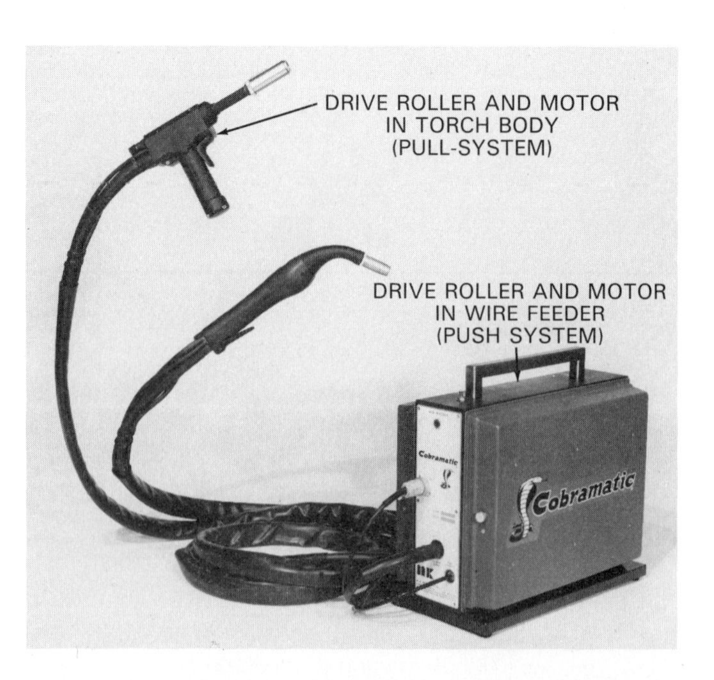

Fig. 3-10. Push-pull systems work extremely well with soft wires even when moving the wire a considerable distance from the wire feeder. (M&K Products, Inc.)

the wire which assists the slave feeder in moving the wire to the work area.

Wire Feeder Controls

The number and types of controls included in the wire feeder will vary depending on the use of the feeder and the amount of wire feed desired.

The major controls, Fig. 3-11, will include:

1. Off/On switch.
2. Wire feed potentiometer (speed control).
3. Spool brake control (stops wire spool at end of welding).

The auxiliary controls, Fig. 3-11, may include:

1. MODE SWITCH. This control is used when the machine may be used in various modes such as spot, stitch, and seam welding.
2. TRIGGER LOCKIN CONTROL. This control allows the welder to weld without the operating switch being depressed during the entire operation.
3. BURNBACK CONTROL. (Antistick Control.) This control sets the time that the arc power is on after the stop switch is released to burn back the wire from the molten pool. This prevents the wire from sticking into the weld pool. This control must be used when spot welding.
4. SPOT WELD TIME. This controls the length of time for the spot weld operation.
5. STITCH WELD TIME. This controls the time needed to make a long seam or a specific weld length.
6. PREPURGE TIMER CONTROL. This control sets the time that the gas will flow before welding will start.
7. PURGE CONTROL. This control allows the welder to open the gas solenoid. The amount of gas flow can be set on the flowmeter or the welder can purge the torch gas lines prior to welding.
8. WIRE INCHING CONTROL. (Wire jog control.) The wire can be moved out of the wire feeder to the contact tip or it may be used by the welder when determining wire speed. (Jog speed is the same speed as the set wire speed.)

Wire Feeder Drive Roller System

The system for driving the electrode wire from the spool to the gun is either a two roller or a four roller system. A standard two roller system is shown in Fig. 3-12, and a four roller drive system is shown in Fig. 3-13. The roller shown in Fig. 3-14 is connected to the drive motor and is grooved for driving a specific size and type of wire. If the groove does not match the wire size, the wire will seize in the groove and not feed properly.

The upper part of the system is the idler roller which is smooth and without grooves. This roller must be adjusted to feed the wire without slippage or flattening of the wire into the groove.

Inlet guides are used to feed the wire into the drive rollers. Outlet guides are also used to feed the wire from the drive rollers into the torch cable or cable adapter. All of the guides are made of a variety of materials. The hard wires (steel, stainless steel, Inconels, etc.) will wear the guides quite rapidly and require replacement often. They are adjusted as close to the roller as possi-

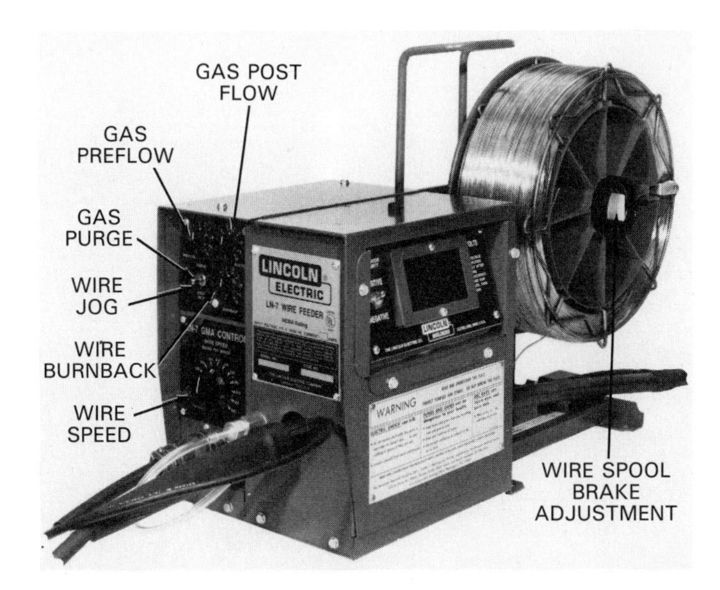

Fig. 3-11. Manufacturers place varying controls on wire feeders depending on the use of the machine. (Lincoln Electric Col.)

Fig.3-12. Two-roll wire feed system. Note the wire grooves in the feed roll. This roller can adapt to two different sizes of wires. (CK Systematics)

ble. This prevents the wire from "birdnesting" in the area between the rollers and the guides which is shown in Fig. 3-15.

Wire Wipers

This accessory is attached to the inlet side of the drive roll system as shown in Fig. 3-16. It usually consists of a piece of felt material that is closed where the wire passes through it. Add a little oil to the felt to assist in the lubrication of the wire. This will extend the life of the guides and the cable liners in the torch cable. Do not over oil and use only when welding on steel materials.

Wire Feeder Maintenance

Wire feeders require very little maintenance to keep them operating properly. As with any electrical device, they must be kept dry to protect the electrical components within the feeder. Electrical malfunctions should be checked within the feeder only by a skilled electrician. The printed circuit boards within the feeder (if so equipped) should be replaced only with factory authorized parts which may be obtained from the welding supplier.

The major problems will be in the drive roller system and the guides to and from the rollers. Worn equipment will cause many problems in the system. Inspect the guides and rollers often and replace when worn. If you have feeding problems with the wire feed,

Fig. 3-13. Four roll wire feed system. These systems are used where precision drive of the filler wire is required. These rollers have the wire size stamped on the outside of the roller. (Miller Electric Co.)

Fig. 3-15. "Birdnesting" of the wire is caused by improper operation of the wire feed roller system. Check the guides, rollers, and tension to find the problem and correct this condition.

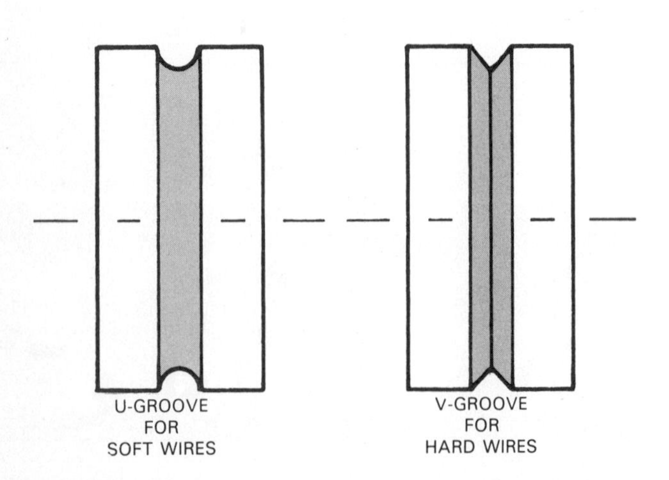

Fig. 3-14. Wire drive rollers have different design grooves for hard or soft wires. Using incorrect grooves causes wire feeding problems.

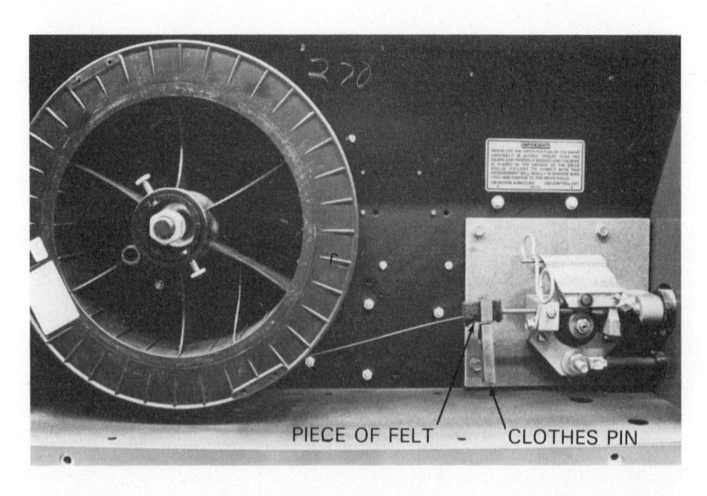

Fig. 3-16. A wiper assembly can be made from a clothespin and a piece of felt. Commercial fluids or light grade oil may be used in the wiper material. Do not over oil.

chances are this is where the problem exists. To determine if the feeder is at fault, disconnect the torch cable and run the wire through the feeder only. The wire should run through smoothly. If the feeder is operating satisfactorily, connect the torch and cable to the feeder. Operate the feeder with the torch and cable attached. If the fault remains, the problem is in the torch and cable assembly. Be sure you know where the problem is before you try to repair it.

CABLES, TORCHES, AND GUNS

Cables

Cables are used to carry electrical current, gas, welding wire, and, in some cases, water in and out to cool the torch. Some cables will also have a circuit wire from the switch to the machine contactor which is used to start and stop the weld operation. The cables are usually made in specified length sections. Some may be connected to other sections lengthening the cable.

Where the cable attaches to the wire feeder, special adapters are necessary. As a general rule, each type of torch requires an adapter which is usually made by the manufacturer of the torch. A cable adapter is shown in Fig. 3-17.

To protect the torch cable from wear as the wire passes through it, a liner is installed into the cable. This liner is made for hard and soft wires and is made to fit specific sizes of wire. Installation of the liner requires that the liner fit into the torch tip adapter with a specific dimension. See Fig. 3-18. Each manufacturer will list the various liners for the different wires in the instruction manual. Since each manufacturer makes torches to a specific design, the liners are not interchangeable.

Fig. 3-17. Different types of torches may be connected to wire feeder with an adapter. (Air Reduction Co.)

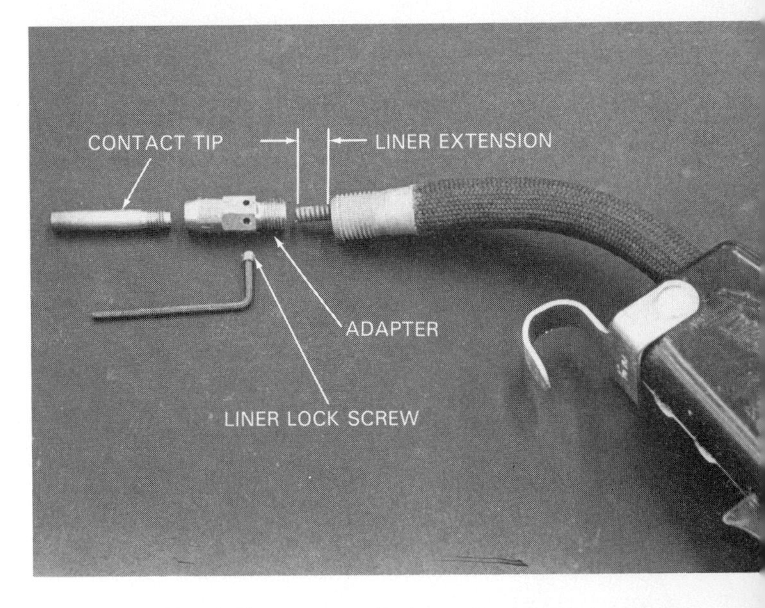

Fig. 3-18. Torch liners must fit into the adapter with a specific dimension. Incorrect assembly of this part will cause wire feeding problems. Always follow manufacturers' instructions when installing liners. (Twecco Products, Inc.)

Torches

Torches are designed to carry current with a specific duty cycle, just as welding machines are. They are usually rated by the type of gas that is being used when they are gas cooled. Torches that are used for large diameter wires are usually water cooled for a higher duty cycle. A water cooled torch with a water cooler mounted in the system is shown in Fig. 3-19.

If city water is used for cooling as shown in Fig. 3-20, the system requires a filter and a pressure regulator to regulate the pressure to the manufacturer's requirements.

These additions are needed because city water usually contains small particles which will clog torch cooling passages and ruin the welding torch. For this reason,

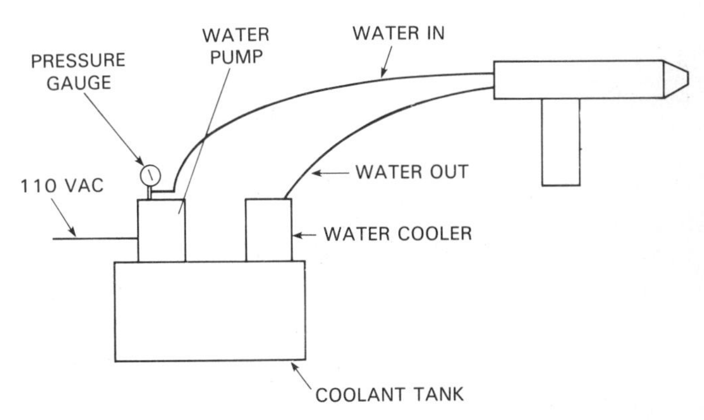

Fig. 3-19. This type of a unit is called a closed system and removes many problems associated with water cooling of the torch.

a large filter is required which must be cleaned often or the filter assembly must be replaced to prevent reduced water flow. A regulator must be installed to prevent too much pressure in the cooling hoses of the torch. Manufacturers usually rate cooling water pressure to be no more than 50 psi in the torch hoses. Since city water pressure varies, the regulator controls the pressure. Failure to regulate the water pressure and filter the water will result in clogged torches, burst hoses, and possibly damage to the welding torch.

Guns

Guns use for welding are rated for duty cycle just as torches are. They contain a drive motor, as shown in Fig. 3-21, which operates on 24 volts direct current and may contain the wire feed adjustment for the wire speed. At this low voltage, they are safe to use without fear of electrical shock. They may be gas or water cooled depending on the duty cycle of the unit.

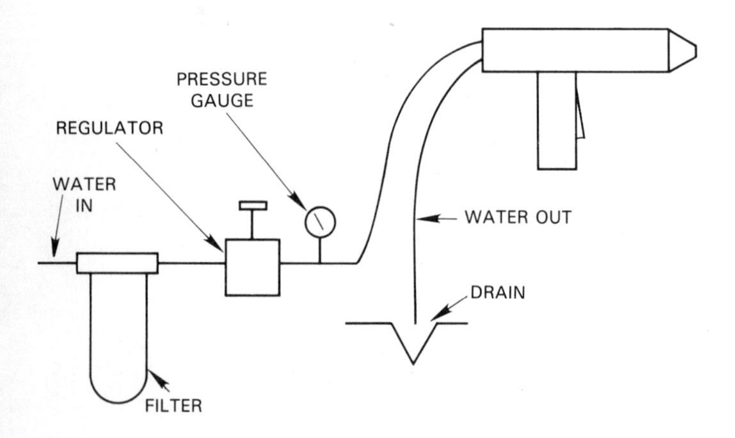

Fig. 3-20. City water systems require more equipment to do an efficient job of cooling without waste of the cooling water.

Torch Parts

As shown in Fig. 3-22, the end of the torch has several components for receiving the liner assembly, insulation of the nozzle, and transferring electrical current to either the contact tip or the electrode. These components should fit together firmly for proper operation of the torch. Insulators which are cracked or burned will not insulate properly and should be replaced.

Contact tips are designed for operation with a certain type and diameter of filler wire and welding mode. When the hole size enlarges to a point where electrical contact is intermittent, the contact tip must be replaced. Continued use of an oversize hole in the contact tip will cause arc outages and current surges to the wire.

Each torch manufacturer lists the proper tip to use for each application with the appropriate part number in the instruction manual. Since these tips are expendable, it is recommended that several tips for each application be kept in stock.

Gas nozzles are made by each manufacturer to fit an individual torch, and they are generally not interchangeable. However, each manufacturer will use the same system for dimensioning the size of the exit diameter. The diameter of the exit nozzle is usually dimensioned in sixteenths of an inch. This is shown in Fig. 3-23. In some cases where high welding currents are carried, the gas nozzle will get extremely hot causing the copper to peel or flake. To assist in the cooling, special aluminum fins are placed on the outer diameter of the nozzle. A special gas nozzle with these fins is shown in Fig. 3-24.

During the welding operation, spatter from the molten metal will gather on the inside of the nozzle which will decrease and deflect the gas flow over the molten metal. To prevent this condition, use a antispatter material inside the nozzle. A compound is shown in Fig. 3-25.

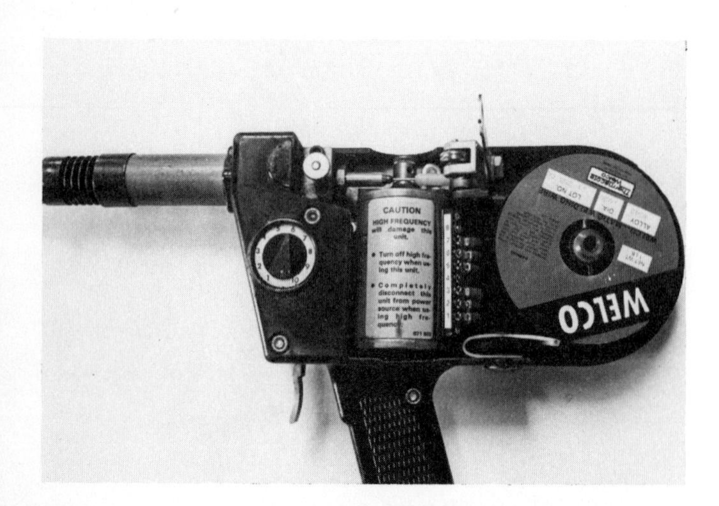

Fig. 3-21. The direct current motor operates on 24 volts and is safe for the operator to use without fear of shock. (Miller Electric Co.)

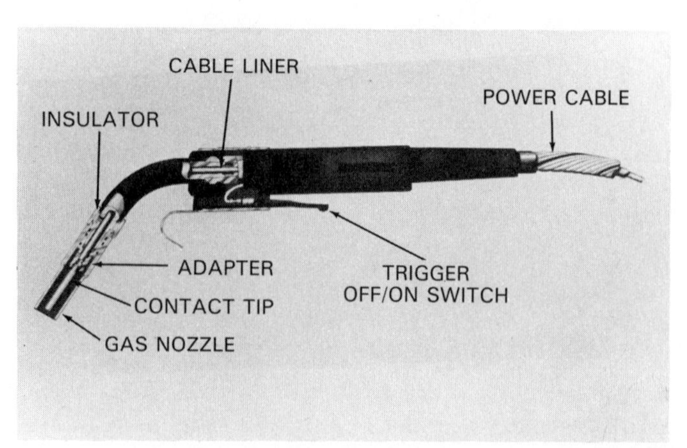

Fig. 3-22. The location of the various components of this torch are shown in the cut-away view. (L-Tech)

SIZE		NO.
3/16	=	NO. 3
4/16	=	NO. 4
5/16	=	NO. 5
6/16	=	NO. 6
7/16	=	NO. 7
8/16	=	NO. 8
9/16	=	NO. 9
10/16	=	NO. 10
11/16	=	NO. 11
12/16	=	NO. 12

LENGTHS*
SHORT
REGULAR
LONG
EXTRA LONG
SPECIAL

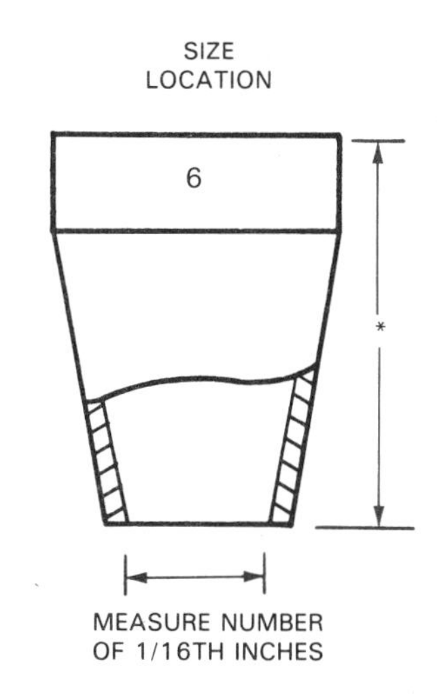

Fig. 3-23. Standard nozzle diameters and lengths.

Fig. 3-24. Air cooled nozzles are used to carry away heat from the nozzle end during high amperage welding. (M&K Products, Inc.)

Fig. 3-25. Anti-spatter spray is used to prevent spatter build up on the end of the nozzle. This should be applied often to the end of the nozzle and the contact tip. (G.S. Parsons Co.)

Torch Maintenance

The welding torch is designed to operate properly provided it is maintained with reasonable care. Treating the torch in a rough manner will damage the outer case and possibly cause internal shortages of electrical current which may ruin the torch. Do not force threaded assemblies together or use parts designed for other torches. When changing adapters or contact tips, use wire brushes to clean threaded areas to assure electrical flow. Check liner set screws often and use compressed gas or air to blow out the liner assembly. Torches or guns which use plastic or steel liners in the front of the torch assembly should be inspected occasionally for wear and possible replacement.

Torches which use a pull system or a wire feeder with segmented teeth should be checked often for proper tension on the wire. The adjustment screw should be tight enough to pull the wire without excessive marking of the wire with the teeth of the wheel. Excessive marking or indentations will cause flaking of the filler wire and rapid wear of the front liner of the torch.

Ground Clamps (Work Leads)

The importance of proper ground leads and proper clamps cannot be overemphasized in GMAW. The cable must be of a sufficient size to adequately carry the welding current without overheating. The ground clamp must fit tightly to the workpiece or an arc will be made when the current flows through the connection. Arc outages may occur where good contact is not made and this part of the weld may not be satisfactory. Cables that have been cracked or broken should be replaced. Connections between the clamp and the cable should be tight. If they are loose, current will not flow properly through the connection. To check the ground system, hold your hand near the cable and connections. If they are warm or hot, they require servicing or replacement. This is one of the major reasons for faulty welding with the GMAW short-arc mode of welding. Several types of ground clamps are shown in Fig. 3-26.

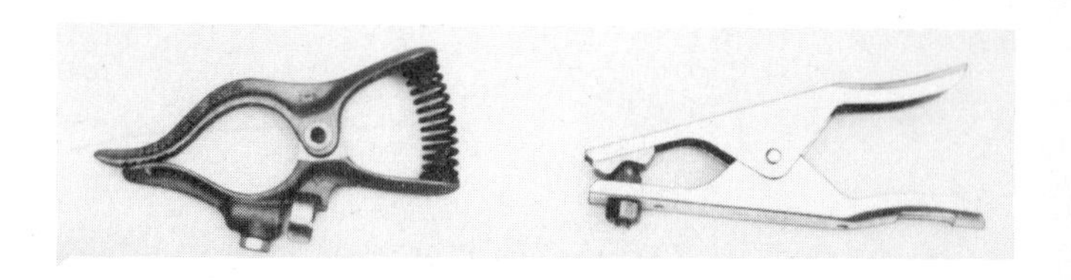

Fig. 3-26. Many different types of ground clamps are available for use with this process. They must work properly and have clean contact points with the part being welded. (Twecco Products, Inc.)

REVIEW QUESTIONS—Chapter 3

1. The power supply produces welding current at ________ voltages.
2. Welding power supplies are made to operate on ________ or ________ phase primary current.
3. All welding power supplies have a ________ output current which is established by the design of the machine to ________ specification.
4. The time at rated load at which the machine will operate without harm to the machine is called the machine ________ ________.
5. This time period is based on a ________ minute period.
6. The letters OCV with regard to welding power supplies define the ________ ________ ________ established on the machine.
7. The actual arc voltage established on the machine after welding begins may also be called ________ voltage.
8. When welding with the short-arc mode, the machine OCV is generally set in the ________ ________ range while the spray-arc mode requires setting on the ________ voltage range.
9. The ________ of the volt-ampere curve is actually the operating characteristics of the welding machine output.
10. When setting up a machine for short-arc welding and slope adjustment is available, the slope is set in the ________ setting.
11. When setting up a machine for spray-arc welding and slope adjustment is available, the slope is set in the ________ mode.
12. Another name for inductance is ________ ________ and it relates to the rate of rise of welding current when the electrode is in contact with the workpiece in the short-arc mode.
13. Adjustment of the inductance control aids in obtaining ________, reducing ________, and obtaining a ________ weld crown.
14. The word ________ is used to express the numbers of cycles of input electrical current.
15. Wire feeders generally use 110 volts input current while gun motors will have a ________ voltage for the safety of the welder.
16. The three types of wire feeders include the ________ type, the ________ type, and the ________-________ type.
17. Wire feeders have either a ________ or ________ wheel drive system. The ________ roll system is used where precision drive of the filler wire is required.
18. Welding machine manufacturers' guarantee on equipment is called the machine ________.
19. Proper ________ on welding equipment will extend the life of the equipment for a long period of time and reduce the number of major repairs.
20. The torch cable is connected to the wire feeder by the use of an ________. This equipment is usually made by the torch manufacturer and only fits the manufacturer's equipment.
21. Torches are rated by ________ ________ just as the power supplies are.
22. The electrical current produced by the power supply is transferred to the electrode through the ________ ________.
23. A ________ is used in the torch cable to protect the cable from wear and is made to fit specific ________ of wire. They are installed in the cable with specific instructions for assembly into the torch adapter.
24. Welding guns which use the push-pull system use ________ ________ wheels that pull the wire from the wire feeder slave unit.
25. Arc outages in the short-arc mode are generally caused by a faulty ________ connection.

Chapter 4

SHIELDING GASES AND REGULATION EQUIPMENT

After studying this chapter, you will be able to:

- Explain what functions shielding gases serve in GMAW.
- Compare characteristics of welds made with different gases or gas combinations.
- Tell which gases or gas combinations are used on ferrous and nonferrous metals.
- Find leaks in a gas distribution system.
- Discuss how shielding gases are supplied and distributed in a GMAW system.

SHIELDING GASES

Shielding gases, either as a single gas or mixed together, are used in GMAW to:

1. Shield the electrode and the molten metal from the atmosphere.
2. Transfer heat from the electrode to the metal.
3. Stabilize the arc pattern.
4. Aid in controlling bead contour and penetration.
5. Assist in metal transfer of the electrode.
6. Assist in the cleaning action of the joint and provide wetting action.

The gases include: argon, helium, carbon dioxide, and oxygen. Argon and helium are inert gases (chemically inactive and will not combine with any product of the weld area). They are pure, colorless, and tasteless and may be used as a single gas or a part of a mixed gas combination. Carbon dioxide is not an inert gas. However, it may be used as a primary gas or a part of a mixed gas combination. Oxygen is always used as part of a mixed gas combination.

Argon

Argon is the most commonly used shielding gas for GMAW. It is separated from the atmosphere during the production of oxygen and is, therefore, readily available at low cost.

Argon produces narrow bead widths, because the arc is more concentrated than that of any other gas. The weld penetration is deep in the center of the weld. Spatter and contamination are reduced since the gas is heavier than air and tends to form a blanket around the electrode and molten metal.

Helium

Helium is found in natural gas wells and is higher in cost than argon. It is not often used as a single gas in GMAW because of poor metal transfer from the electrode to the weld pool. Although helium has a higher thermal conductivity than argon, penetration is wider and not as deep. Since helium is lighter than air, the gas tends to rise from the weld area quite rapidly. This makes a higher flow rate necessary for proper shielding.

Carbon Dioxide

Carbon dioxide is a compound gas made up of carbon monoxide and oxygen. These individual gases are mixed and then stored as a liquid. GMAW uses only carbon dioxide which has had the moisture removed during processing. It is then termed WELDING GRADE CARBON DIOXIDE.

Oxygen

Oxygen is acquired from the atmosphere and is not used as a single gas for welding. It is always used as a part of a gas mixture to attain specific arc patterns. Since the amount used is a very small percentage of the total gas mixture, the user should purchase the gas mixture from the supplier as a mixed gas.

Gases and Mixes for Welding Ferrous Metals

1. 100 percent carbon dioxide may be used for short-arc welding steel.
2. 99 percent argon with 1 percent oxygen may only

be used for welding stainless steel in the spray-arc mode in the flat and horizontal positions. The addition of oxygen to argon assists in stabilizing the arc while making appearance of the weld bead better. Groove welding is limited to the flat position. Fillet welds may be done in either flat or horizontal positions.

3. 98 percent argon with 2 percent oxygen may be used in the spray-arc mode for welding carbon steels and stainless steels. Groove welding is limited to the flat position. Fillet welds may be done in either flat or horizontal positions.
4. 95 percent argon with 5 percent oxygen may be used for spray-arc welding carbon and stainless steels. Groove welding is limited to the flat position. Fillet welds may be done in the flat and horizontal positions.
5. 75 percent argon with 25 percent carbon dioxide is the most common mixture for short-arc welding carbon steels in all positions. Welds made with this gas mixture have minimum spatter and good penetration features.
6. 50 percent argon with 50 percent carbon dioxide offers many qualities of the 75 percent argon with 25 percent carbon dioxide gas mixture only at a lower cost. This is because less of the more expensive argon is used. When spatter and a deeper penetration can be tolerated, this gas mixture may be used.

Gases and Mixes for Welding Nonferrous Metals

1. Argon is used for spray-arc welding in all positions on aluminum, nickel-base alloys, and reactive metals. It may also be used for welding some thin gauge materials with the short-arc mode.
2. Use of helium gas alone is limited in GMAW. Its basic application is for machine welding of heavy aluminum at high welding currents.
3. 75 percent argon with 25 percent helium is normally used on heavier materials with the spray-arc mode. Penetration is deeper than with pure argon and the weld bead appearance is good.
4. 75 percent helium with 25 percent argon is used with spray-arc welding of heavier materials. This percentage of helium increases the heat input which reduces internal porosity and provides good wetting of the weld into the parent metal.

Special Gas Mixtures

1. 90 percent helium with 7 1/2 percent argon and 2 1/2 percent carbon dioxide is a mixture developed for short-arc welding of stainless steels. Welding may be done in all positions. The argon addition provides good arc stability and penetration. The high helium content provides heat input to overcome the sluggish nature of the weld puddle in stainless steel.
2. 60 percent helium with 35 percent argon and 5 percent carbon dioxide is a mixture for short-arc welding of high strength steel in all positions.

Specialty Gas Mixtures

Some special gas mixtures have been developed by gas suppliers for use in standard and special applications. These mixes often allow a broader range of welding variables than the standard mixes and, therefore, allow reduction of welding cost. Some of the gases are named: STARGON, MIG-MIX, etc.

Selecting The Proper Gas Mixture

Selecting the proper gas or gas mixture requires consideration of several factors:

1. Mode of metal transfer.
2. Base metal type.
3. Base metal thickness.
4. Joint design.
5. Weld position.
6. Filler material composition.
7. Filler material size.
8. Chemical composition of the desired weld metal.
9. Weld metal quality.

Fig. 4-1 lists the various gases used in GMAW and the proper applications.

Purge Gas Applications

When purging of the root side of the weld is required for protection from atmospheric contamination, always use either argon or helium. Where an inexpensive gas is desired for initial purging of tanks, vessels, or pipe lines, nitrogen gas may be used prior to the admittance of an inert gas. The gas may also be used to protect the root side of fillet welds on ferrous metals. This type of joint design does not allow admittance of nitrogen into the torch gas. The uses of nitrogen as a purging gas or backing gas are shown in Fig. 4-2.

Gas Purity

Inert shielding gases for welding are refined to high purity specifications. Cylinder argon has a minimum purity of 99.996 percent and contains a maximum of about 15 parts per million moisture (a dew point temperature of −73°F maximum). Driox (Linde Co.) argon has a minimum purity of 99.998 percent and a moisture content of less than six parts per million. Helium is produced to a minimum purity of 99.995 percent. Generally, helium contains less than 15 parts of moisture. At these purity levels, impurities usually cannot be detected during welding.

Steels and copper alloys have high tolerances for various amounts of contaminants. Aluminum and magnesium are sensitive to the gas purity level and will have severe porosity if contaminated gas is used.

Still others, such as reactive metals, have extremely

SHORT-ARC WELDING

METAL	ARGON	HELIUM	ARGON/ HELIUM	ARGON/ CO_2	ARGON/ HELIUM/ CO_2	CO_2
Aluminum	*****	*****	***** (HE-75)			
Carbon steel				***** (C-25) or (C-50)		*****
High strength steels					***** (A-415)	
Copper			***** (HE-75)			
Stainless steels					***** (Tri-mix)	
Nickel alloys	*****	*****	***** (90-HE 10-AR) or (HE-75)		***** (Tri-mix)	
Reactive metals	*****	*****	***** (HE-75 25-AR)			

SPRAY-ARC WELDING

METAL	ARGON	ARGON/ HELIUM	ARGON/ 1% OXYGEN	ARGON/ 2% OXYGEN	ARGON/ 5% OXYGEN
Aluminum	*****	***** (HE-75 AR-25)			
Carbon steel			*****	*****	***** (Plate)
Copper	*****	*****			
Stainless steel			*****		***** (Plate)
Nickel alloys	*****	*****			
Reactive metals	*****	*****			
Silicon bronze	*****				

Fig. 4-1. GMAW gas mixes and applications. (L-Tech)

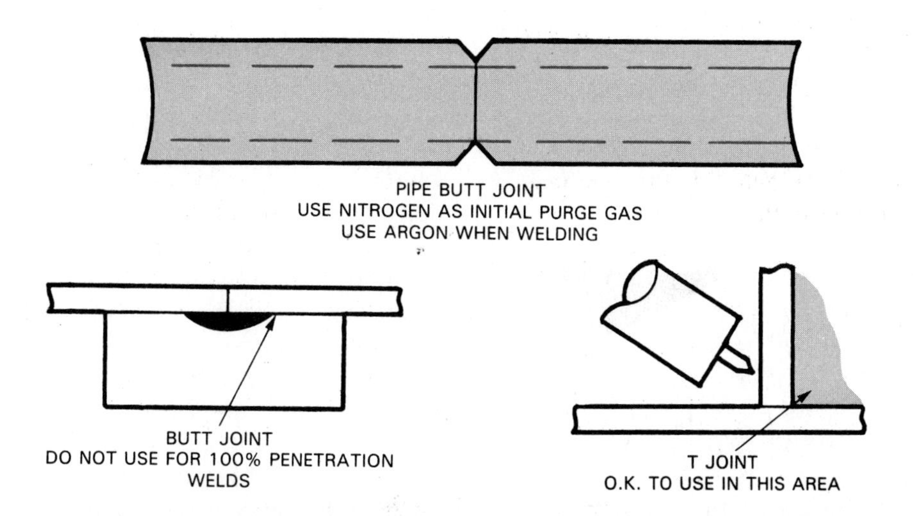

Fig. 4-2. Nitrogen used as a purge gas. (Use only on ferrous metals).

low tolerances for contaminants in the inert gases. Therefore, high purity standards are maintained by the gas suppliers to insure the shielding gases used will be more than adequate for the most severe application.

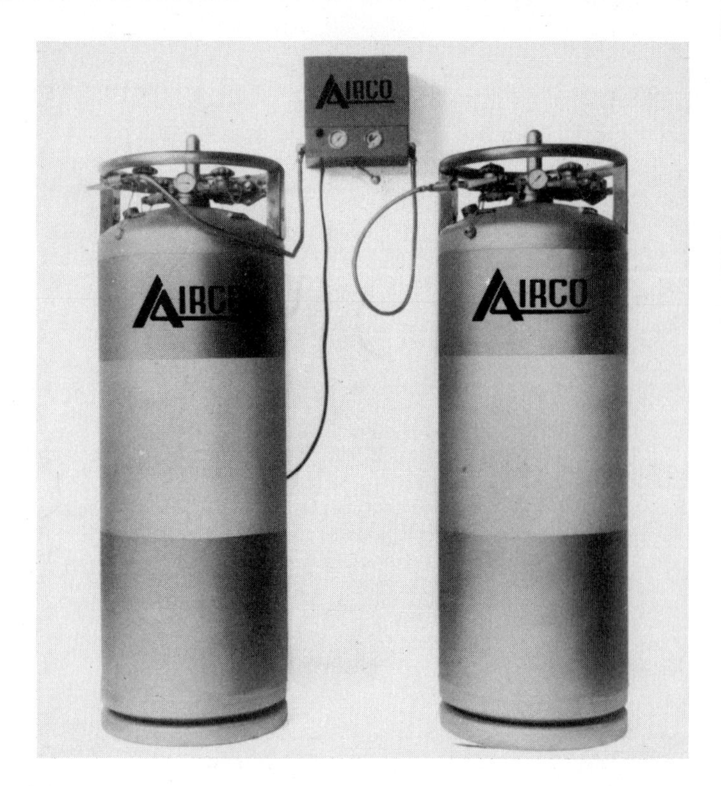

Fig. 4-4. Two argon Dewars are connected to a pressure regulator and a switching valve. (Air Reduction Co.)

Gas Supply

As presented in Fig. 4-3, shielding gases are supplied in cylinders of various sizes for shop use. Dewars (liquefied gas container), as shown in Fig. 4-4, and high pressure gas cylinders mounted on trailers are used where high volumes of gases are required.

In most cases, the gas is sold through a distributor by the total number of cubic feet of each type of gas. As the gas is distributed in cylinders, the distributor will also charge a demurrage (rental fee) on each cylinder used. In areas where large amounts of gases are used, liquefied gases are the most economical because fewer cylinders are used.

Storage

Storage of gas cylinders and containers should be rigidly controlled to prevent the incorrect use of the shielding gases. The gas cylinder or container should always be stored in an outside or well ventilated area. It should also be in an upright position while secured to a rigid support. The safety cap should be used for all cylinders not in use. The cylinder should be properly identified as to type or mixture of the gas contents. All safety precautions should be followed to avoid injury to the user. Remember, inert gases do not contain oxygen, and therefore, will not support life. You cannot see, smell, or taste inert gases.

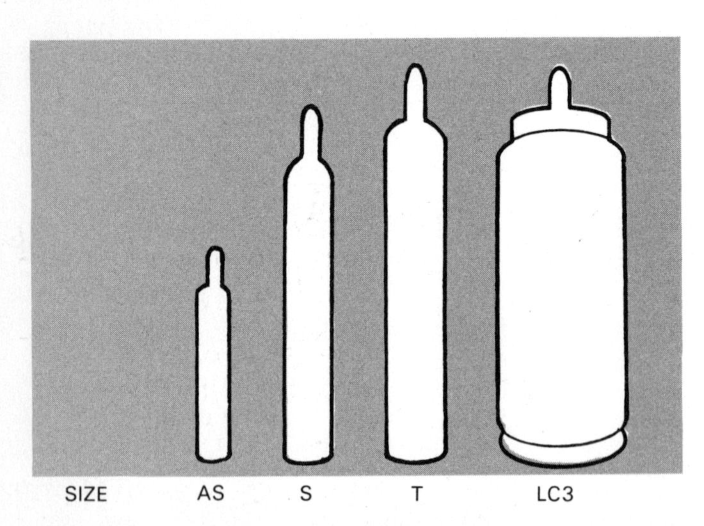

CYLINDER STYLE	CONTENTS CUBIC FT.	FULL PRESSURE OF CYLINDER AT 70 °F	HEIGHT	O.D.
AS	78	2200	35	7 1/8
S	150	2200	51	7 3/8
T	330	2640	60	9 1/4
LC-3	2900	55	58	20

Fig. 4-3. Types and capacities of cylinders and Dewars supplied to industry.

Gas Distribution

Gases may be distributed to the welding area in several ways. Fig. 4-5 shows a bank of cylinders that have been connected together. They may be located in a convenient place with the gas being piped to the welding area. With this arrangement, one bank may be used until empty, and then another can be placed into use. The empty cylinders, in turn, can be replaced without affecting the bank in use.

Manifolds are often used to distribute gases to the welding area from the supply area. Through the use of manifolds, the number of individual cylinders required at the welding station can be reduced. The manifold in Fig. 4-6 can be used to supply up to six stations. Fig. 4-7 shows a commercial manifold connecting several banks together.

The distribution system must be leak free to maintain the purity of the gas being used. Therefore, all of the cylinder fittings must be cleaned before installation and properly seated into the regulators. High pressure connectors, tubing, and pipe connectors must be protected to prevent entry of foreign materials, water, or oil when the system is not in use. Plastic thread adapters or tape may be placed over any unused openings for this purpose.

Fig. 4-5. Individual cylinders are connected to a supply manifold by high pressure tubing. These tubes are often called "pig tails." (Air Reduction Co.)

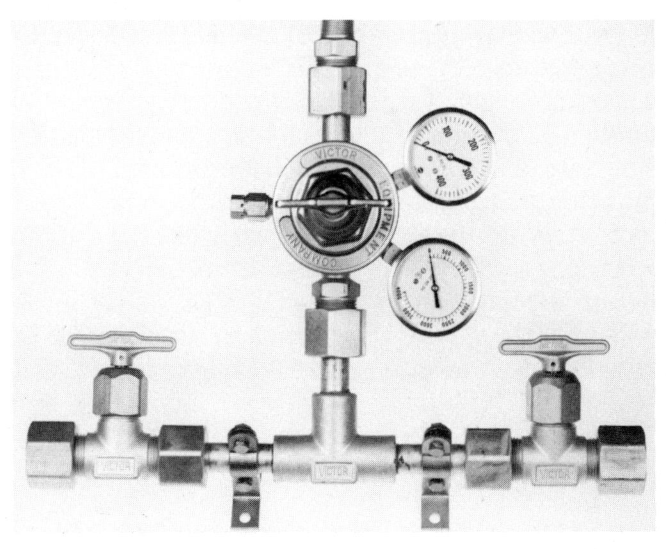

Fig. 4-7. This manifold is used for high pressure gases. A regulator is used to reduce the cylinder pressure to the desired manifold pressure. (Victor Equipment Co.)

Testing for Gas Leaks

Before placing a system into use, it must be tested for leaks at a pressure above the normal operating pressure. One method of testing is to use a special leak test solution, Fig. 4-8. While the pipe is under pressure, the solution is applied. Since it is sensitive to gas flow, the solution forms bubbles wherever a leak is located.

Another method for testing is to apply pressure to the system and note the test pressure. The inlet pressure valve is then closed. By observing the pressure gauge, one will know a leak is present by a drop in pressure. It should then be repaired.

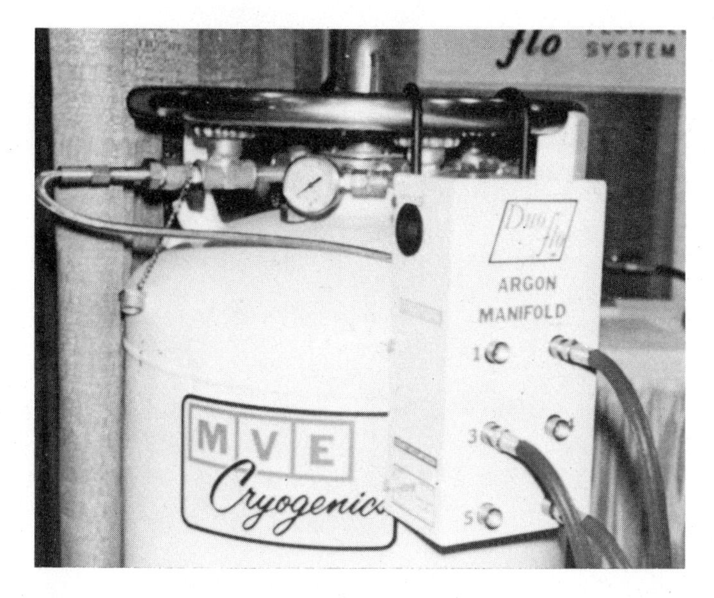

Fig. 4-6. Dewars supply gas at approximately 50 psi. A six station manifold is attached to this Dewar. (Distribution Designs, Inc.)

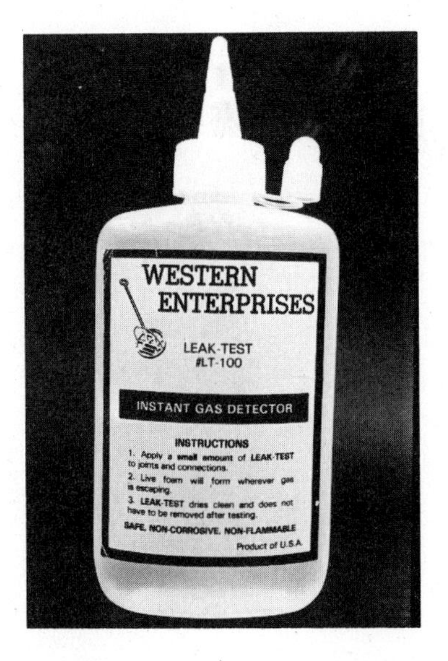

Fig. 4-8. The leak test solution is applied to the pressurized joint. The solution will bubble if a leak is present. (Western Enterprises Co.)

Prior to placing the manifold in use, the lines must be purged to remove any flux vapors and moisture. This can be done by using nitrogen gas as the initial purging gas. Connect the gas to the manifold, set the flow rate on the other end of the pipe to 5 CFH and purge until analyzer tests show the system is clear of oxygen. When the test shows clear, connect the argon to the line and purge at the same rate for several hours before welding.

Should the system require repair, disassembly, or modification, the entire system should be repurged and tested prior to use.

Gas Regulation

Regulation of shielding gases is accomplished by the use of several types of special equipment. Gases distributed through a manifold require regulators, Fig. 4-7. They reduce the pressure from the cylinder to the desired manifold pressure level. This pressure is usually set from 20 to 50 psi. A flowmeter like the one in Fig. 4-9, is then used at the welding station. It regulates the flow of gas to the welding torch.

Where a cylinder is used at the welding station, the regulator/flowmeter is used to reduce the pressure from the cylinder and regulate flow to the torch. See Fig. 4-10 and Fig. 4-11.

When installing a regulator/flowmeter or a flowmeter with a ball tube, the ball tube should always be in the vertical position for proper operation. The amount of flow is indicated at the top of the ball unless otherwise indicated.

Regardless of the type of gas supply (cylinder, Dewar, manifold), when the gas flow valve is opened, a surge of gas will exit from the gas nozzle. This is due to the pressure buildup when the gas is not flowing. This surge of gas will last for several seconds until the excess pressure is reduced. To eliminate this condition, a specially designed surge check valve may be used, as shown in Fig. 4-12.

Fig. 4-9. Station flowmeters of this type operate on manifolds at approximately 50 psi. (Linde Co.)

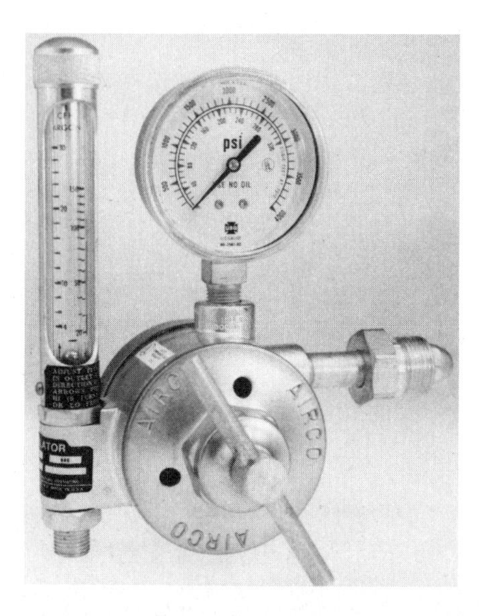

Fig. 4-10. Single cylinder regulator/flowmeters have a gauge that shows cylinder pressure. Gas flow (cubic feet per hour) to the torch is adjusted by turning the adjustment knob and reading the ball position in the vertical tube. (Air Reduction Co.)

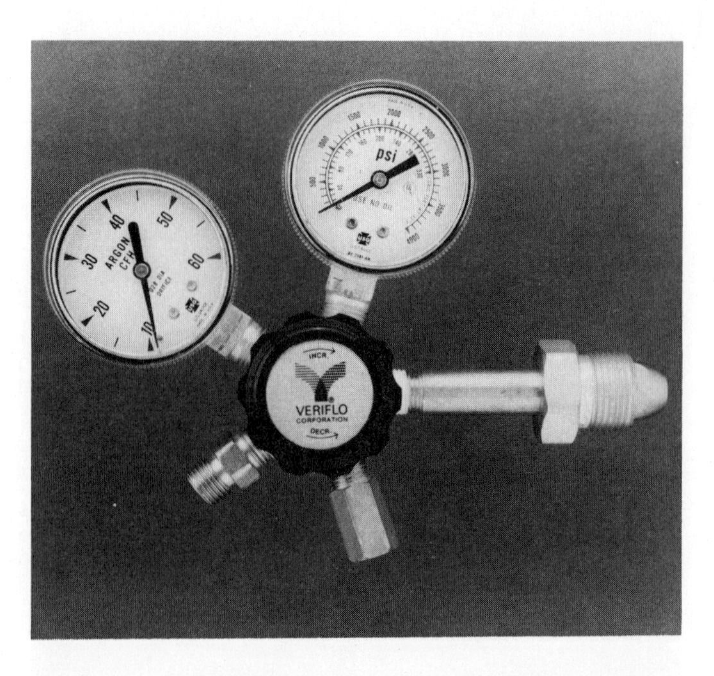

Fig. 4-11. A dial gauge is used with a regulator to read gas flow in cubic feet per hour. The desired gas flow is obtained by turning the adjustment knob on the front of the regulator. (Veriflow Co.)

Gas Mixing

Welding gases can be mixed at the manifold, as in Fig. 4-13. This type of mixer is helpful where large volumes of gases are needed. The gas mixer in Fig. 4-14 is used on single station installations. The mixer in Fig. 4-15 may be employed in single or multiple station installations. Another type of mixer, pictured in Fig. 4-16, may also be used in single stations with the gas mixed within the Y valve arrangement. These Y valves are installed on the outlet side of the flowmeter with the gas metered by two separate gas flowmeters.

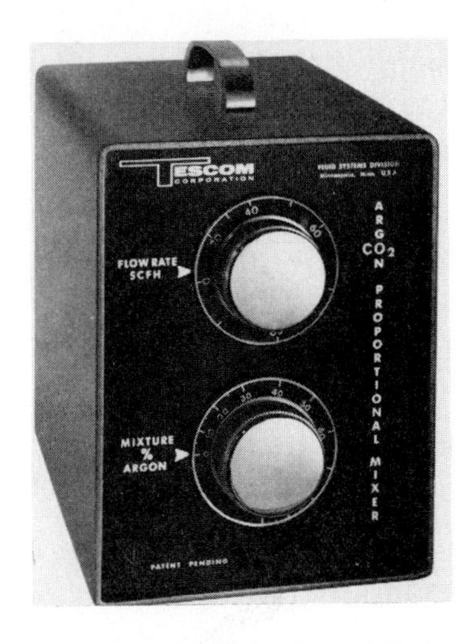

Fig. 4-14. Small proportional mixers of this type are used for individual station mixing. (Tescom Corp.)

Fig. 4-12. Surge check valves eliminate surging of gas from the nozzle during the start of the operation. By eliminating the surge of gas they readily pay for themselves in gas savings. (Weld World Co.)

Fig. 4-15. Mini-proportional mixers with storage tanks are used for individual or manifold stations. (Air Reduction Co.)

Fig. 4-13. The large tank on the bottom of the mixer serves as a mixing tank and storage chamber. This assures sufficient volumes of mixed gases for large users. (Thermco Instrument Co.)

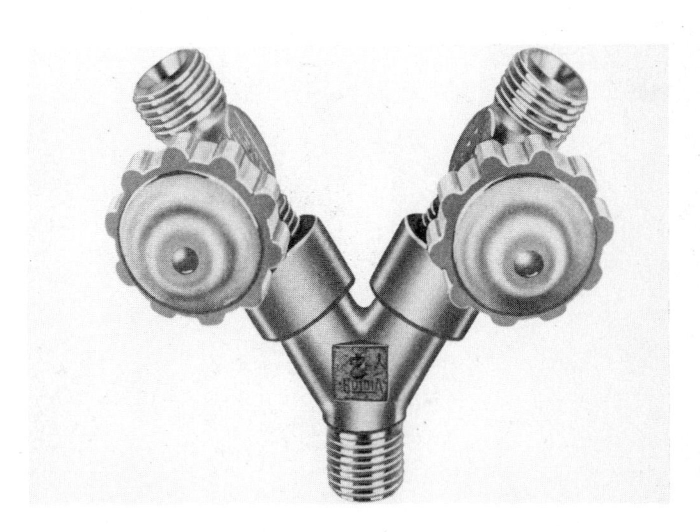

Fig. 4-16. Y valves have shut off valves for the single flow of one gas or a mix of two gases. (Victor Equipment Co.)

To prevent back flow of the gases and improper mixing, a flow check valve, as shown in Fig. 4-17, should be installed between the flowmeter and the Y valve. Fig. 4-18 shows a back flow check valve installed on a flow meter.

Mixture and Purity Testing

Gas analyzers test for proper mixes at the welding stations, Fig. 4-19. These instrument can also be used for leak checking and testing for adequate purging of pipes, vessels, etc., prior to welding.

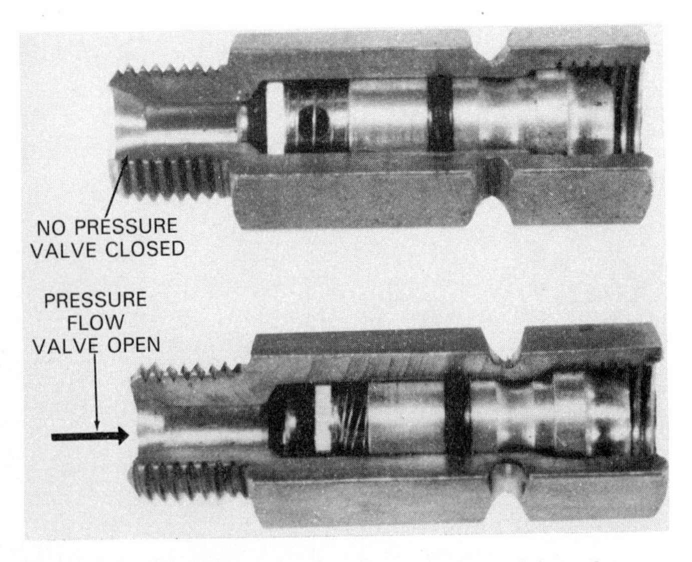

Fig. 4-17. Back flow check valves prevent mixing of gases in the supply line. The upper valve is shown closed. The bottom valve is shown open. (Air Reduction Co.)

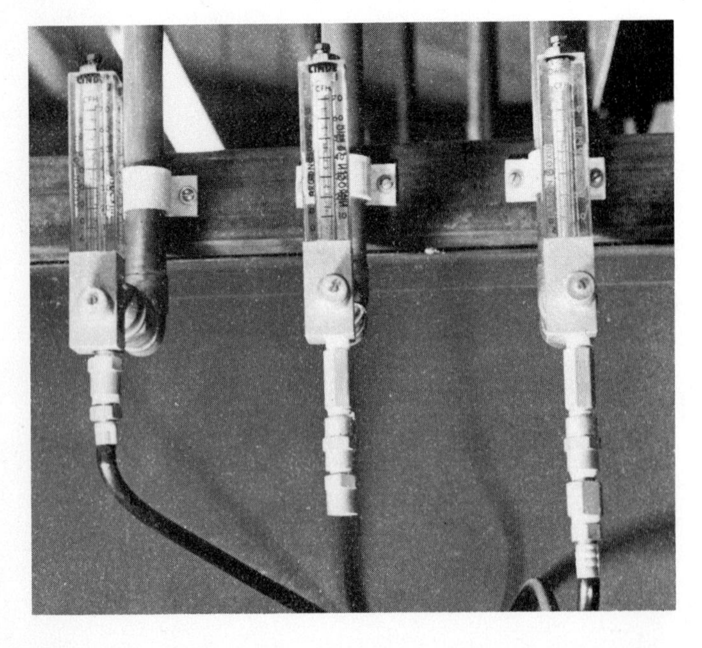

Fig. 4-18. Back flow check valves are installed on the outlet side of the flowmeter.

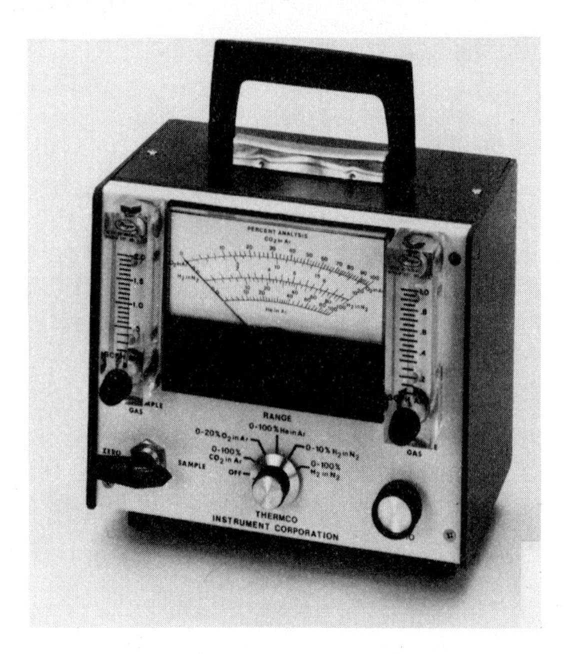

Fig. 4-19. Portable analyzers of this type measure the percentages of the individual gases. (Thermco Instrument Co.)

REVIEW QUESTIONS—Chapter 4

1. Gases are used in GMAW to:
   A. ________ the electrode and the molten pool from the atmosphere.
   B. ________ heat from the electrode to the metal.
   C. ________ the arc pattern.
   D. ________ in controlling bead contour and penetration.
   E. ________ in metal transfer of the electrode.
   F. ________ in the cleaning action of the base material.
2. Two inert gases used in the GMAW process are ________ and ________.
3. Two gases used in the GMAW process which are not inert are ________ and ________ ________.
4. ________ is always used as a part of a gas combination.
5. Argon gas is ________ than air and tends to form a blanket around the molten pool.
6. ________ gas is lighter than air and requires higher flow rates than argon.
7. Carbon dioxide is a manufactured gas made by combining ________ ________ and ________ together.
8. Only ________ ________ carbon dioxide can be used in the GMAW process.
9. When used alone, ________ gas is primarily used on nonferrous materials.
10. When used alone, ________ ________ gas is primarily used on carbon steels.

11. When used in the GMAW process, ________ gas is limited to purging applications.
12. Gas purity is defined as moisture content in ________ ________ ________.
13. Dewars are containers filled with ________ ________.
14. Charges made by a gas supplier for the use of gas cylinders are called ________.
15. Gas cylinders in storage should always have the ________ ________ installed until ready for use.
16. Cylinders should always be stored and used in the ________ ________ and ________ to a solid object to prevent accidental tipping and falling of the cylinder.
17. Cylinder gas must always be reduced from the tank pressure with a ________ ________.
18. The gas flow used during welding is metered by a ________ to register flow rates in ________ ________ ________ ________ or ________ ________ ________.
19. Shielding gases can be checked for gas purity or gas percentages with a ________ ________.

Chapter 5

FILLER MATERIALS

After studying this chapter, you will be able to:

- ■Describe how filler materials are made.
- ■Compare the application of general use, rigid control use, and critical use filler wire.
- ■Tell how contamination of filler wire is avoided by both the manufacturer and the welder.

Filler materials used in the GMAW process must be of the highest quality to make acceptable welds. For this purpose, manufacturers of welding wire use specialized machines, processes, and inspections. They supply filler wire in rolled form to fit various types of wire feeders.

Manufacturing

Material which is to be made into filler wire is selected on several factors:

1. Chemical composition.
2. Mechanical properties.
3. Notch toughness values.
4. Impurity level limits.

Material selected at the primary mill is hot drawn through reducing dies to predetermined sizes, then cleaned. Further reduction of the wire is then done cold. Annealing (softening) and cleaning operations are performed as the wire is further reduced in size.

Various types of lubricants are used in the draw dies during the drawing process. The lubricants simultaneously reduce wear of the dies and decrease friction between the wire and the dies. Lubricants also carry away heat produced in the dies during the drawing operation.

During the drawing process, several types of defects can occur which will affect the quality of filler materials. These defects include:

1. Overlapping.
2. Splitting.
3. Cracking.
4. Seaming.
5. Oxide formations.

Examples are shown in Fig. 5-1. Filler materials with these defects should not be used. After the drawing, another process cleans the wire of all surface impurities. Then the wire is processed for shipment to the user.

Specifications

Inspectors use many quality specifications for the manufacturing, testing, inspecting, and packaging of filler materials. Fig. 5-2 lists some of the common specifications. These may include the following requirements:

1. Scope.
2. Classification and usability.
3. Manufacturing methods.
4. Acceptability tests.

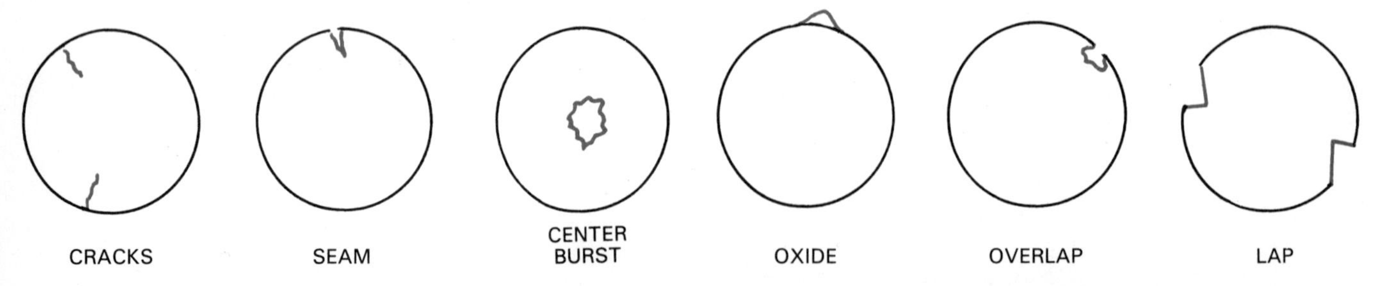

Fig. 5-1. Typical defects that can occur in the wire during manufacturing.

Specifications	Material Types
American Welding Society	All Types
American Society for Testing Materials	All Types
Mil-R-5632	Steel
Mil-R-5031	Stainless Steel
Mil-E-23765	Steel
Mil-E-16053	Aluminum
QQ-R-566	Aluminum
Aeronautical Material Specifications	All Types

Fig. 5-2. Common welding wire specifications used by the welding industry to establish quality.

5. Chemical composition.
6. Usability tests and results.
7. Standard sizes.
8. Finish and temper.
9. Spool and winding requirements.
10. Packaging.
11. Marking of packages.
12. Guarantee.

In some cases, the user may add some additional requirements to the basic specification such as chemistry, packaging, and weld testing. Each additional requirement will add cost to the filler material.

Manufacturers of filler materials only guarantee their product to meet specification. This means they will only replace defective wire. They do not guarantee acceptable results, as they cannot govern the welding application.

Filler wire manufacturers make welding material for three major areas:

1. GENERAL USE. This wire will meet specification requirements. However, no record of chemical composition, strength level, etc., is submitted to the user when the wire is purchased.
2. Fabrication that requires rigid control over the filler material. Wire used under this condition may require a CERTIFICATE OF CONFORMANCE with the purchase of the material. This certificate, shown in Fig. 5-3, is a statement that the filler materials meet all of the requirements of the material specification. All of the stock will be identified by heat numbers, lot numbers, or code numbers located on the wire roll package. On some work, the buyer may require that these numbers are recorded wherever the material is used in welding applications.
3. CRITICAL USE. Welding operations, such as on aircraft, nuclear reactors, and pressure vessels, usually require very close control of the filler material's chemistry. A CERTIFIED CHEMICAL ANALYSIS report, like the one in Fig. 5-4, is the

CERTIFICATION OF QUALITY CONFORMANCE TESTS

Manufacturer or Distributor ________ Customer's Name ________

Address ________

Date ________ Customer's Order No. ________

Specification MIL- ________

Type MIL- ________ Core Wire Heat No. ________

Diameter & Length ________ Lot Identification MIL- Para. ________

Inspection Level ________

Lot No. ________ Wet Batch No. ________

Chemical Analysis (Complete)

Carbon ________

Manganese ________

Silicon ________

Phosphorus ________

Sulfur ________

Chromium ________

Nickel ________

Molybdenum ________

Vanadium ________

Mechanical Test	As-Welded	Stress-Relieved
Yield Strength (0.2% offset method)	____	____
Tensile Strength	____	____
Elongation (%)	____	____
Reduction of Area (%)	____	____

Charpy Impacts

1. ____	1. ____
2. ____	2. ____
3. ____	3. ____
4. ____	4. ____
5. ____	5. ____

Chemistry was taken from: Chem Pad ☐ Groove Weld ☐

X-Ray Results ________

Concentricity (%) ________

Covering Moisture ________

Grinding During 8a Test Plate Preparation 3/

Groove Weld Test

Test No. 3 8a Chem Pad

Amperage ____ ____ ____

Operator Error (Layer Nos.) ________

We hereby certify that the above material has been tested in accordance with the listed specification and is in conformance with all requirements.

____

Fig. 5-3. Certificate of conformance form. (Techalloy Maryland, Inc.)

Reid - Avery

Dundalk, Baltimore, Md. 21222

QUALITY ASSURANCE TEST REPORT

DATE: ________

SOLD TO: SHIPPED TO: DATE SHIPPED: ________

P.O. NO.:- P.O. NO.:-

SPECIFICATION:

ITEM	POUNDS	SIZE	TYPE	LOT NO	HEAT NO
1.					
2.					
3.					

ACTUAL CHEMICAL ANALYSIS OF WIRE OR WELD METAL

ITEM	C	Mn	P	S	Si	Cr	Ni	Mo				
1.												
2.												
3.												

ADDITIONAL TEST RESULTS

TESTS PERFORMED WITH: ________

X-RAY: ________

STRESS RELIEVED ________ HRS. @ ________ °F

AS WELDED PLATE NO. ________ STRESS RELIEVED PLATE NO. ________

YIELD POINT, PSI

TENSILE STRENGTH, PSI

ELONGATION IN 2", %

REDUCTION OF AREA, %

IMPACTS PLATE NO. ________ @ ________ °F ________ ft. lbs. PLATE NO. ________ @ ________ °F ________ ft. lbs.

State of ________

City of ________

Subscribed and sworn to before me this ________ day

of ________ 19 ________

Notary Public

My commission expires ________

I certify the chemical analysis and physical or mechanical test results reported above are correct as contained in the records of the company.

QUALITY ASSURANCE DEPARTMENT

Fig. 5-4. Certified chemical analysis form. (Techalloy Maryland, Inc.)

actual chemical analysis of the individual heat or lot of material. The test is made on a spectrometer machine shown in Fig. 5-5. Records are maintained during fabrication cycles wherever the specific materials are used. In case of a joint failure, either because of the filler material or the base metal, other welded joints in the weldment or system can be located and evaluated for possible replacement.

Filler Material Form

Filler materials used in GMAW are wound on spools or inside of large drums depending on the application. Standard spools are 4, 8, or 12 inch diameter. They are made of plastic, wood, or formed metal wire. The spools are disposable after use. See Fig. 5-6.

The wire may be either level wound, Fig. 5-7, or layer wound, Fig. 5-8. This depends on the material type and the application.

Fig. 5-5. Actual chemical analyses are obtained from a wire sample on this machine. (Techalloy Maryland, Inc.)

The wires are furnished in standard sizes which include:

1. .023 inch diameter.
2. .030 inch diameter.

Fig. 5-6. Various types of wire spools which fit standard wire feeders.

Fig. 5-7. The aluminum welding wire on this spool is level layer wound.

Fig. 5-8. The steel welding wire on this spool is layer wound.

3. .035 inch diameter.
4. .045 inch diameter.
5. 3/64 inch diameter (soft wires).
6. .062 inch diameter.

Different types and sizes of wire have varying amounts of wire on the spool. Nonferrous materials will always have less metal per spool than ferrous materials.

To prevent wire feeding problems during welding, the wire must meet cast and helix requirements. CAST is the diamter of one complete circle of wire from the spool, as it lies on a flat surface. With hard wire, this diameter tends to be larger than the diameter on the spool. HELIX is the maximum height of any point of this circle of wire above the flat surface. Cast and helix dimensions and tolerances are shown in Fig. 5-9. These two dimensions are critical to the feeding of filler material to the welding arc. Improper dimensions will affect the entire wire feed system and may cause "bird-nesting," as shown in Fig. 3-15. Other adverse effects are arc outages, severe liner wear, and contact tip wear.

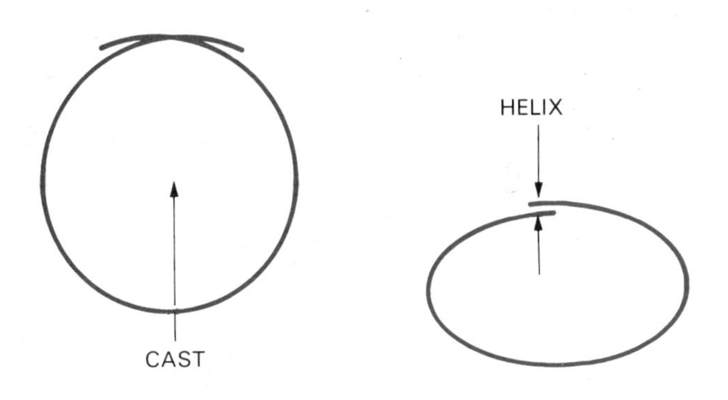

Standards for Cast and Helix

Spool Size	Wire Types	Cast Min. in	Cast Min. mm	Cast Max. in	Cast Max. mm	Helix Max. in	Helix Max. mm
4-in (100 mm)	Low-Alloy, Stainless, and Nickel Alloy	6*	150	9	230	½	13
	Aluminum	†		6	150	1	25
8-in (200 mm)	Low-Alloy, Stainless, and Nickel Alloy	15*	380	30	760	1	25
12-in (300 mm)	Low-Alloy, Stainless, and Nickel Alloy	15*	380	30	760	1	25
	Aluminum	†		15	380	1	25

* Measured on outside strand of full spool
† Diameter of wire level from which sample is taken

Fig. 5-9. Cast and Helix. These terms define the characteristics of any form of continuous wire as it comes from the spool or coil. Cast is the diameter of one complete circle of wire as it lies on a flat surface. Helix is the maximum height of any point of this circle of wire above the flat surface.

Identification

Labels or tapes on the inside of the spool hub or on the flange identify spools or coils of wire. An identifying label is pictured in Fig. 5-10.

Filler Material Packaging

Manufacturers of filler materials use a variety of packaging methods to protect the material during shipment and storage. The spools may have special paper to protect the material, or they may be wrapped in plastic, as shown in Fig. 5-11. Then, they are packaged in cardboard cartons or metal cans.

Fig. 5-10. The label identifies this material as: 4043 aluminum. HQ letters identify the material as a special processed high quality material. The Linde stock number is 5202F07. The Linde control number is 707. The actual heat number is 00350. The wire is 3/64 inch diameter. The weight of the welding wire on the spool is 14.3 pounds. (Linde Co.)

Fig. 5-11. Spooled wire in this package will store indefinitely if the package remains sealed.

Filler Material Use in the Shop

Filler materials are easily contaminated by oil, moisture, grease, soot, and salts from the hands. Dirty gloves, work areas, rags, etc., readily contaminate the filler wire whenever they contact it. These foreign materials often cause defects in welds such as porosity or cracks. Rework of a rejected weld is costly.

Filler materials are packaged to prevent contamination of the material during shipment. Preventing contamination of the material after the package is opened is the responsibility of the user.

The simplest method of avoiding contamination of the filler material is to keep it clean. Use the following practices:

1. Keep the material packaged as long as possible.
2. Open packages only when needed.
3. Store unsealed filler materials in a heated cabinet.
4. Handle material as little as possible, then only with clean gloves.
5. Work in a clean and dry area.
6. Remove spools from machines when machines are to be idle for an extended period of time.

Contaminated filler materials cannot be tolerated where high quality welds are required. Requiring the filler wire manufacturer to clean, inspect, and package the material according to specifications does little good if the material is contaminated prior to use by improper handling. Quality welds cannot be made with contaminated filler metal.

Filler Material Selection

One of the most important factors to consider in GMAW is the correct filler wire selection. This wire, in combination with the shielding gas, will produce the deposit chemistry that determines the final physical and mechanical properties of the weld. The following basic factors influence the choice of filler wire:

1. Base metal chemical composition.
2. Base material mechanical properties.
3. Type of shielding gas used.
4. Service use of the weldment.
5. Type of weld joint design.

The actual selection of filler wire to be used for a particular welding operation is shown in various chapters relating to a specific material.

REVIEW QUESTIONS—Chapter 5

1. Welding wire selected at the mill for drawing into various diameters is based on what four factors?
2. As the wire is drawn through the dies to reduce the diameter, it becomes hard and must be ________ to make it soft.
3. List five types of defects produced during the drawing operation which can make the wire unusable.
4. One of the major specifications used for the purchase of welding wire is made by the ________ ________ ________.
5. ________ ________ wire is obtained for welding where no specifications are defined by the manufacturer of the welded product.
6. What document is used when manufacturing specifications require a record of wire quality?
7. A report of the actual chemical analysis of the material is stated on a ________ ________ ________.
8. Welding wire is wound on spools of ________, ________, or ________ inch diameters.
9. Wires manufactured to a fractional size dimension are considered to be ________ wires.
10. Major problems may be encountered in the feeding of the spooled wire during the welding operation if the ________ and ________ requirements are not correct.
11. ________ and ________ of welding wire is of prime importance in helping prevent contamination of the filler wire.
12. List five basic factors which influence the choice of filler material to be used in welding.

Chapter 6

WELD JOINTS AND WELD TYPES

After studying this chapter, you will be able to:

- ■Identify different types of welds used with various joints.
- ■Read and draw common welding symbols.
- ■Discuss advantages and disadvantages of different weld joints.
- ■List factors involved with joint design.

JOINT TYPES

The American Welding Society defines a JOINT as the manner in which materials fit together. There are five basic types of joints. They include:

1. Butt joint.
2. T joint.
3. Lap joint.
4. Corner joint.
5. Edge joint.

The basic types of joints are shown in Fig. 6-1.

WELD JOINT PREPARATION

Weld joints may be initially prepared in a number of ways. These include:

1. Shearing.
2. Casting.
3. Forging.
4. Machining.
5. Stamping.
6. Filing.
7. Routing.
8. Oxyacetylene cutting (thermal cutting process).
9. Plasma arc cutting (thermal cutting process).
10. Grinding.

Final preparation of the joint prior to welding will be covered in chapters which detail the welding of a particular material.

WELD TYPES

Various types of welds can be made in each of the basic joints. They include:

1. Butt joint, Fig. 6-2.

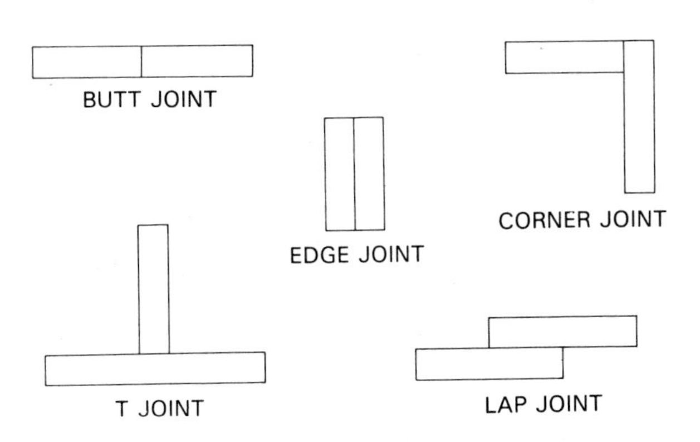

Fig. 6-1. Basic types of joints.

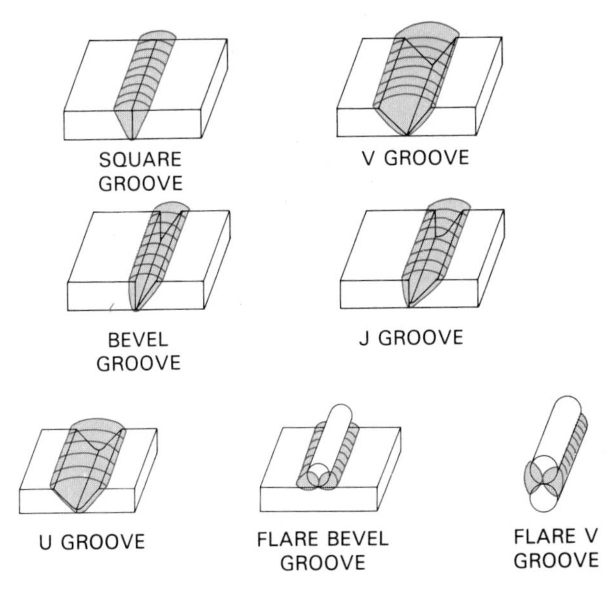

Fig. 6-2. Types of welds which may be made with a basic butt joint.

A. Square groove butt weld.
B. Bevel groove butt weld.
C. V groove butt weld.
D. J groove butt weld.
E. U groove butt weld.
F. Flare V groove butt weld.
G. Flare bevel groove butt weld.

2. T joint, Fig. 6-3.
   A. Fillet weld.
   B. Plug weld.
   C. Slot weld.
   D. Bevel groove weld.
   E. J groove weld.
   F. Flare bevel groove weld.
   G. Melt through weld.
3. Lap joint, Fig. 6-4.
   A. Fillet weld.
   B. Plug weld.
   C. Slot weld.
   D. Spot weld.
   E. Bevel groove weld.
   F. J groove weld.
   G. Flare bevel groove weld.
4. Corner joints, Fig. 6-5.
   A. Fillet weld.
   B. Spot weld.
   C. Square groove weld or butt weld.
   D. V groove weld.
   E. Bevel groove weld.
   F. U groove weld.
   G. J groove weld.
   H. Flare V groove weld.
   I. Edge weld.
   J. Corner flange weld.

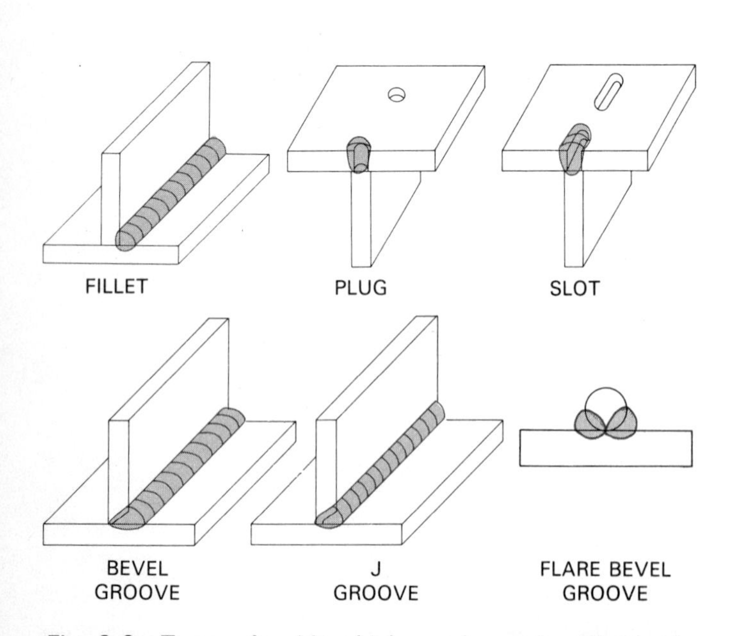

Fig. 6-3. Types of welds which may be made with a basic T joint.

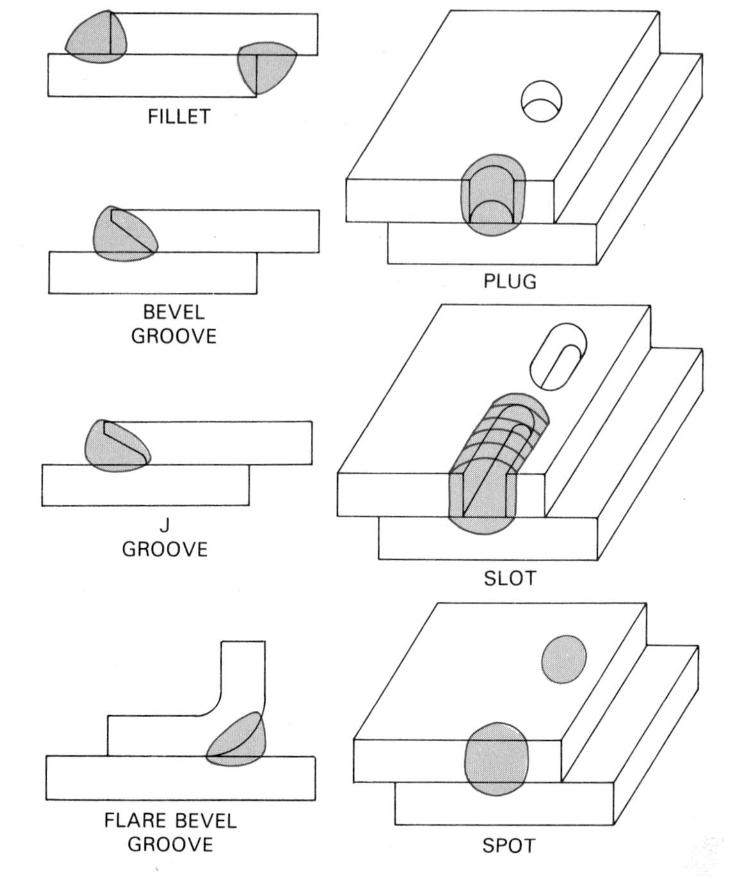

Fig. 6-4. Types of welds which may be made with a basic lap joint.

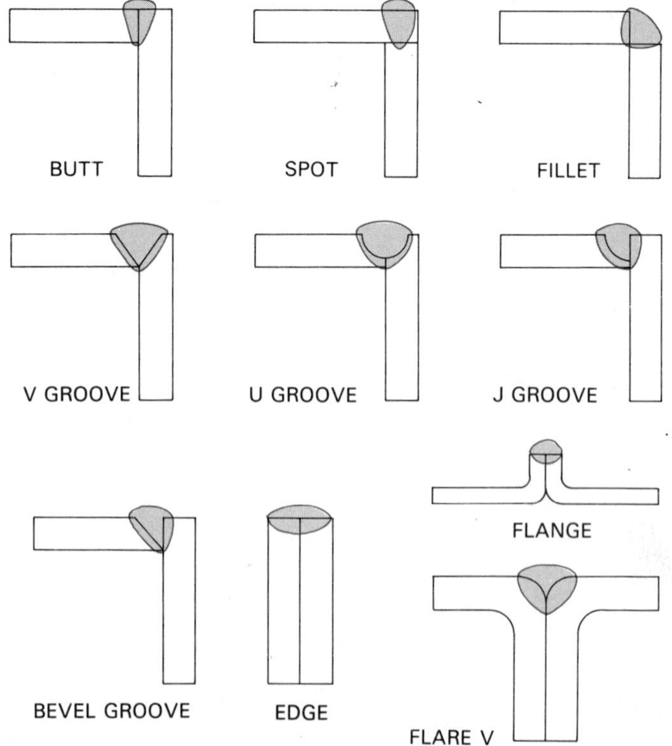

Fig. 6-5. Types of welds which may be made with a basic corner joint.

5. Edge joints, Fig. 6-6.
   A. Square groove weld or butt weld.
   B. Bevel groove weld.
   C. V groove weld.
   D. J groove weld.
   E. U groove weld.
   F. Edge flange weld.
   G. Corner flange weld.

DOUBLE WELDS

In some cases, the weld cannot be made from one side of the joint. When a weld must be made from both sides, it is known as a DOUBLE WELD. Fig. 6-7 shows common usage of double welds on basic joint designs.

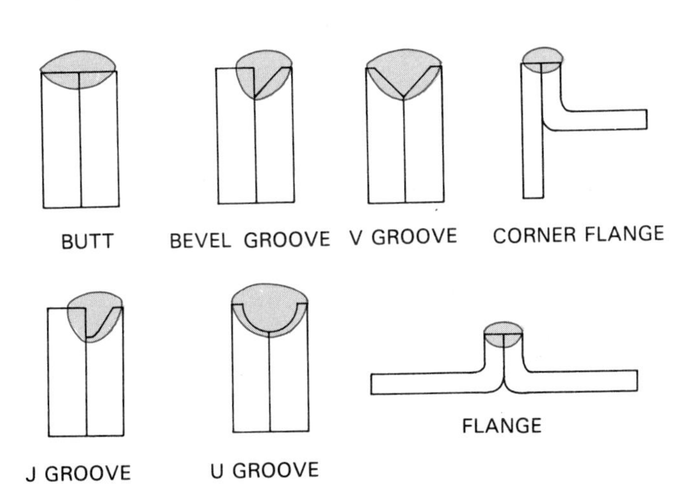

Fig. 6-6. Types of welds which may be made with a basic edge joint.

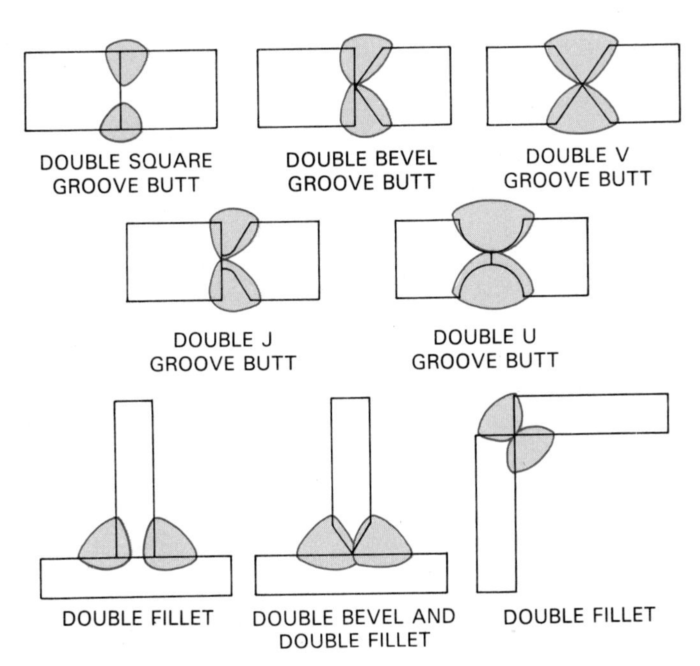

Fig. 6-7. Applications of double welds.

WELDMENT CONFIGURATIONS

Often, the basic joint is changed to assist in a component's assembly. A weld joint might be modified to gain access to the weld joint or to change a weld's metallurgical properties. Some common weldment configuration designs are described here. Joggle type joints are used in cylinder and head assemblies where backup bars or tooling cannot be used, Fig. 6-8. Another application of joggle joints is in the repair of unibody automobiles where skin panels are placed together and welded. Tubing to heavy wall tube is shown in Fig. 6-9. A built-in backup bar is used when enough material is available for machining the required backup. Fig. 6-10 shows a fabricated backup bar. These bars must fit tightly or problems will be encountered in heat flow and penetration. Specially designed weld joints for controlled penetration are used where excessive weld penetration causes a problem with assembly or liquid flow. This type of joint is found in Fig. 6-11.

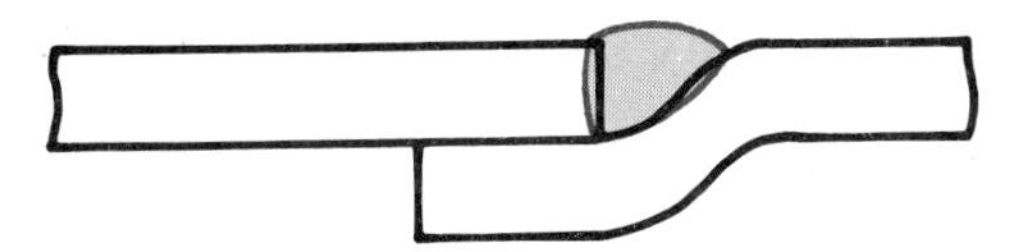

Fig. 6-8. Joggle type joint.

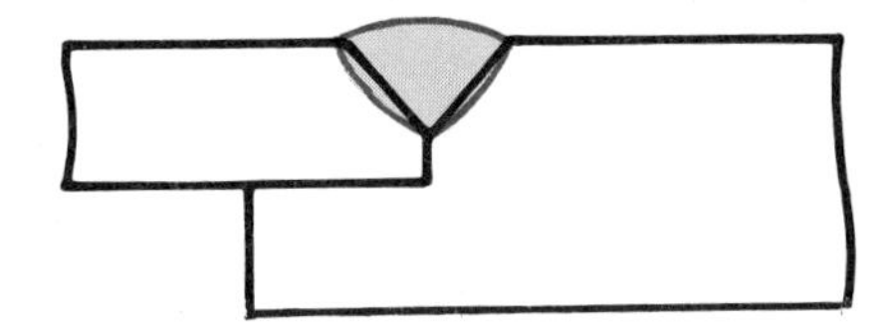

Fig. 6-9. Built-in backup bar joint.

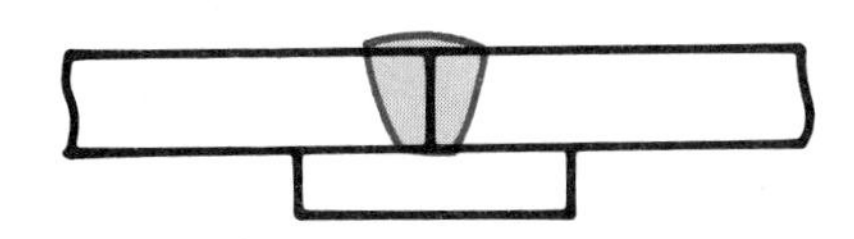

Fig. 6-10. Fabricated backup bar configuration.

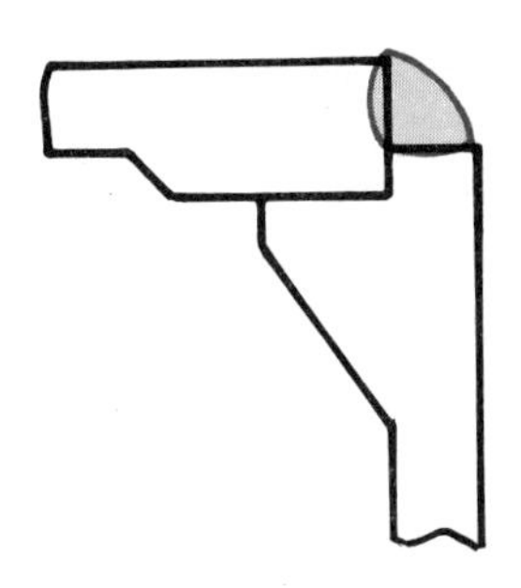

Fig. 6-11. Controlled penetration joint.

A series of bead welds overlayed on the face of a joint is called BUTTERING, Fig. 6-12. Buttered welds are often used to join dissimilar metals. Series of overlayed welds on the surface of a part to protect the base material is called SURFACING or CLADDING.

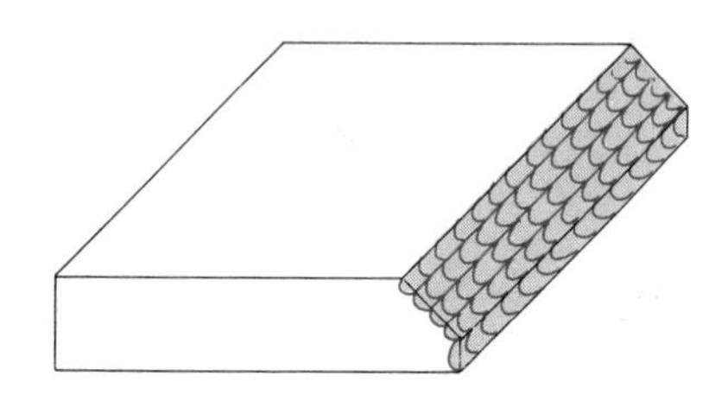

Fig. 6-12. Buttered weld joint face.

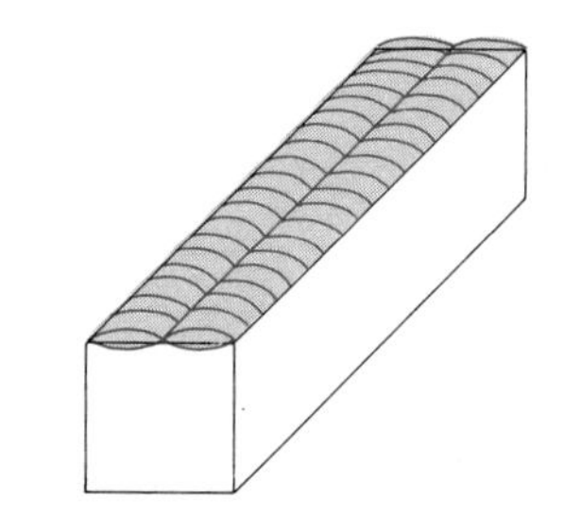

Fig. 6-13. Overlay welds.

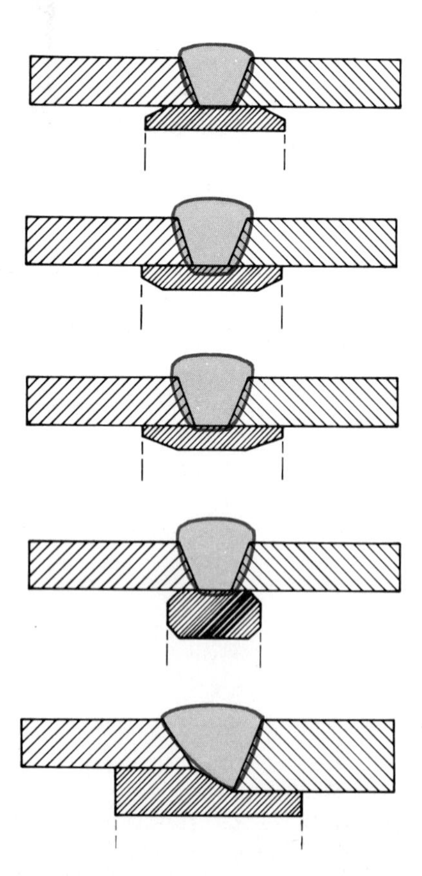

Fig. 6-14. Various types of backup rings for pipe joints.

Refer to Fig. 6-13. Pipe joints often use special backup rings or are machined to fit specially designed mated parts. Types of backup rings are shown in Fig. 6-14.

WELDING TERMS AND SYMBOLS

Communication from the weld designer to the welder is essential in the manufacturing of most weldments. Some common terms used to describe the weld joint are found in Fig. 6-15. Other terms used to describe welds are given in Fig. 6-16. The welding symbol shown in Fig. 6-17 has been adopted by the American Welding Society. This symbol is used on drawings to indicate the type of weld joint, placement, and the type of weld to be made. The symbol may also include other information regarding finish contours and possibly testing information of the completed weld.

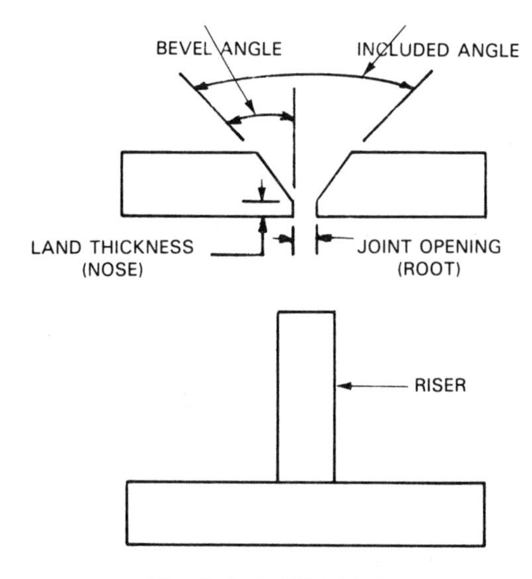

Fig. 6-15. Weld joint terms.

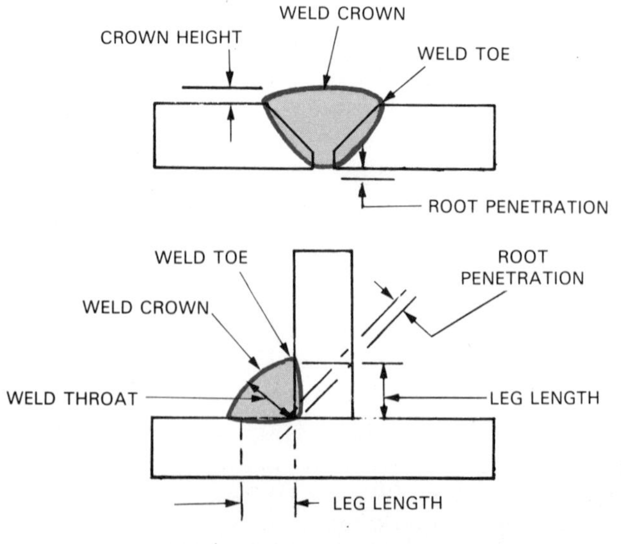

Fig. 6-16. Weld and weld area terms.

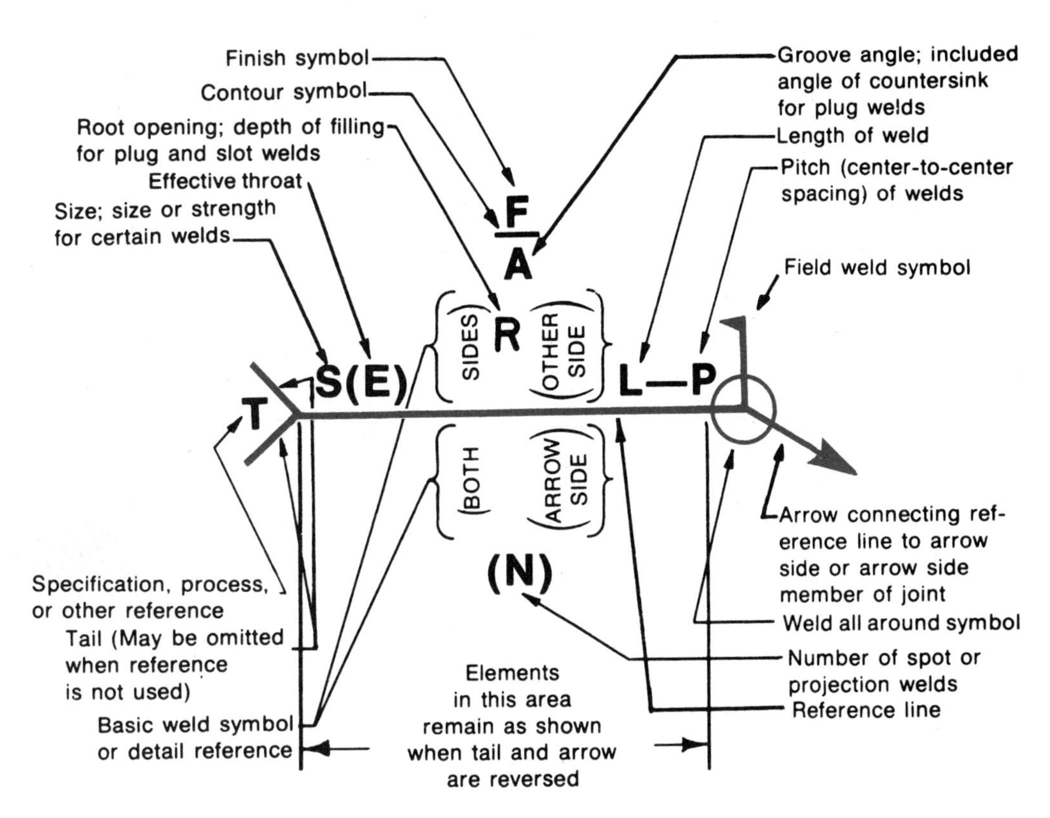

Fig. 6-17. The AWS welding symbol gives complete and specific welding information to the welder.

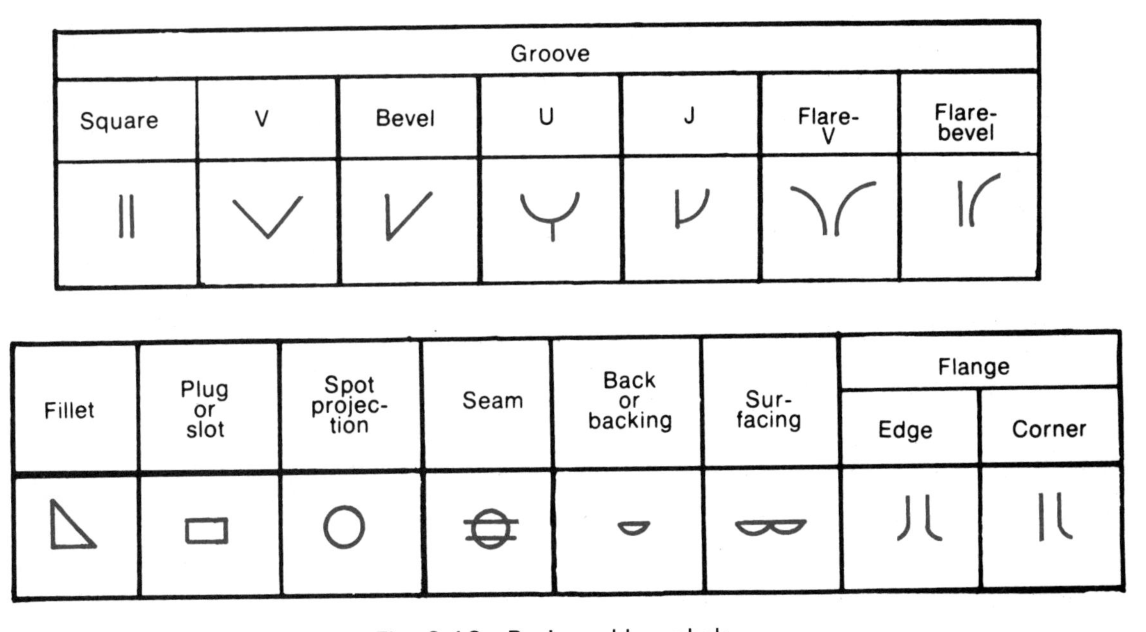

Fig. 6-18. Basic weld symbols.

It is important to study and understand each part of the welding symbol. Fig. 6-18 contains basic weld symbols which direct the welder to select the proper weld joint. The arrow indicates the point at which the weld is to be made. The line to the arrow is always at an angle to the reference line. Whenever the basic weld symbol is placed below the reference line, the weld is made where the arrow points, as shown in Fig. 6-19. Whenever the basic symbol is placed above the reference line, the weld is to be made on the other side of the joint, as shown in Fig. 6-20. By placing dimensions on the symbol and drawings, the exact size of the weld may be indicated. Study the examples of typical weld symbols and weldments shown in Fig. 6-21.

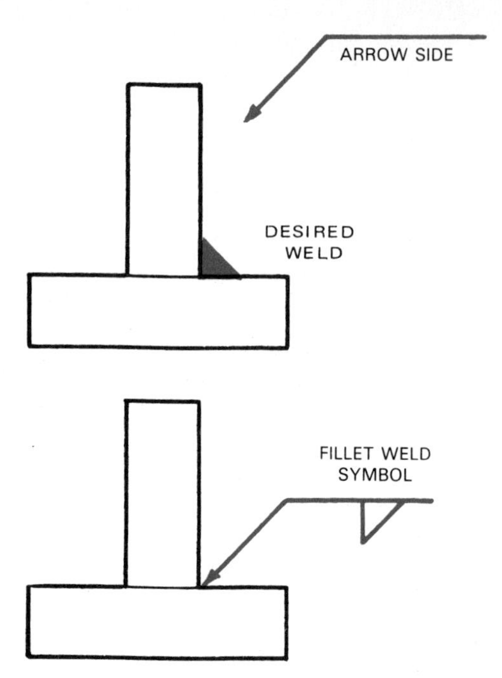

Fig. 6-19. Fillet weld symbol shown on the bottom side of the reference line indicates that the weld is located where the arrow points.

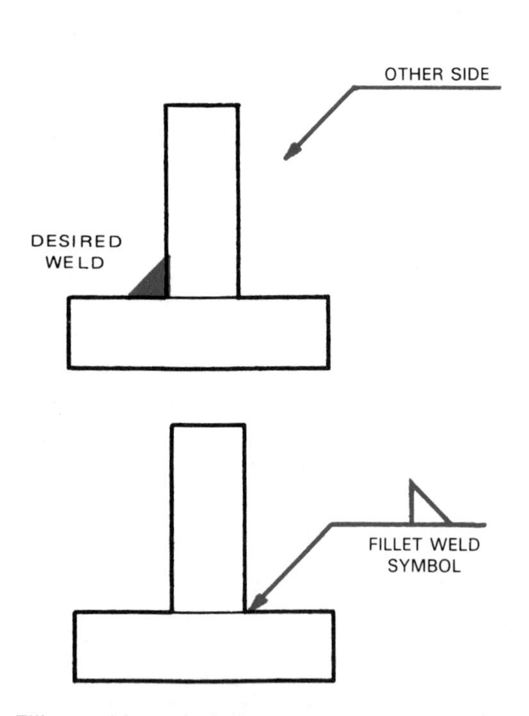

Fig. 6-20. Fillet weld symbol shown on the upper side of the reference line indicates that the weld is located on the opposite side of the joint.

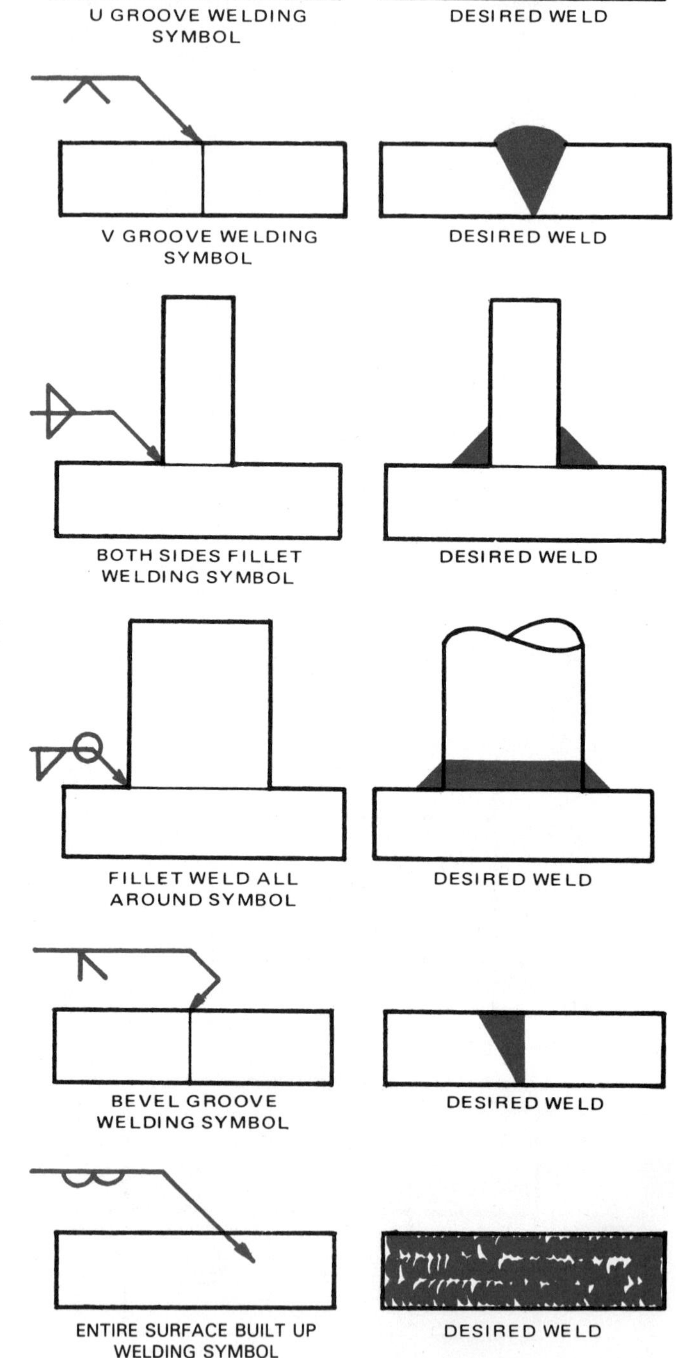

Fig. 6-21. Typical weld symbols and weld applications.

The complete weld symbol gives the welder instructions to prepare the base metal, the welding process to use, and the finish to add to the completed weld. Through careful use of these symbols, the weld designer can convey all the information needed to complete a weldment.

Many classes are offered which provide advanced study in the area of blueprint reading for welders. By taking such classes, the welder can improve his or her ability to read and interpret welding drawings. Another method of gaining ability to read prints is studying texts on blueprint reading.

WELD POSITIONS

For a welder, it is important to be able to weld in different positions. The American Welding Society has defined the positions of welding to include:

1. Flat.
2. Horizontal.
3. Vertical.
4. Overhead.

Fig. 6-22 demonstrates the four positions for fillet welds, grooved butt welds, and pipe welds. While practicing welding in these positions, the student welder should note how gravity affects the molten weld pools. In addition to this, heat distribution also varies with each position. These factors make the skills needed for each position distinct. Practice is required to produce good welds in all positions.

DESIGN CONSIDERATIONS

Design of the weld type and weld joint to be used are of prime importance if the weldment is to do the intended job. The weld should be made at reasonable cost. Several factors concerning the weld design must be considered. Evaluations should be made concerning the areas of:

1. Material type and condition (annealed, hardened, tempered).
2. Service conditions (pressure, chemical, vibration, shock, wear).
3. Physical and mechanical properties of the completed weld and heat affected zone.
4. Preparation and welding cost.
5. Assembly configuration and weld access.
6. Equipment and tooling.

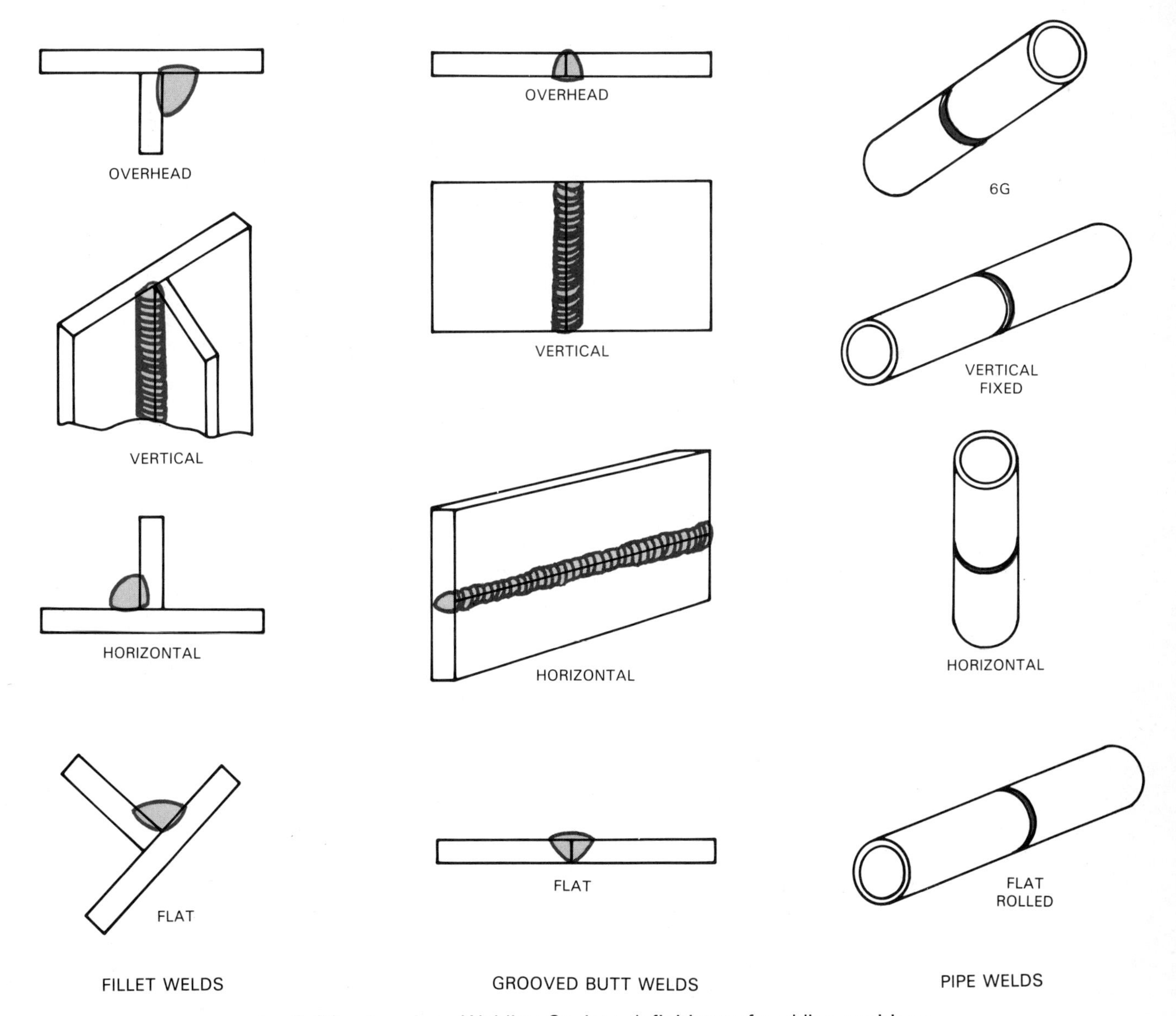

Fig. 6-22. American Welding Society definitions of welding positions.

Butt Joints and Welds

Butt joints are used where high strength is required. They are realiable and can withstand stress better than any other weld type. To achieve full stress value, the weld must have 100 percent penetration through the joint. This can be done by welding completely through from one side. The alternative is welding both sides with the welds joining in the center.

Thinner gauge metals are more difficult to fit up for welding. Thinner metals also require more costly tooling to maintain the proper joint configuration. Tack welding may be used as a method of holding the components during assembly. However, tack welds present many problems, including:

1. They conflict with the final weld penetration into the weld joint.
2. They add to the crown dimension (height).
3. They often crack during welding due to the heat and expansion of the joint.

The expansion of the base metal during welding will often cause a condition known as MISMATCH, as shown in Fig. 6-23. When mismatch occurs, the weld generally will not penetrate completely through the joint. Many specifications limit highly stressed butt joints to a 10 percent maximum mismatch of the joint thickness.

Whenever possible, butt joints should mate at the bottom of the joint, Fig. 6-24. Joints of unequal thicknesses should be tapered in the weld area to prevent incomplete or inadequate fusion. This is shown in Fig. 6-25. Where this cannot be done, the heavier piece may be tapered on the upper part of the joint as well.

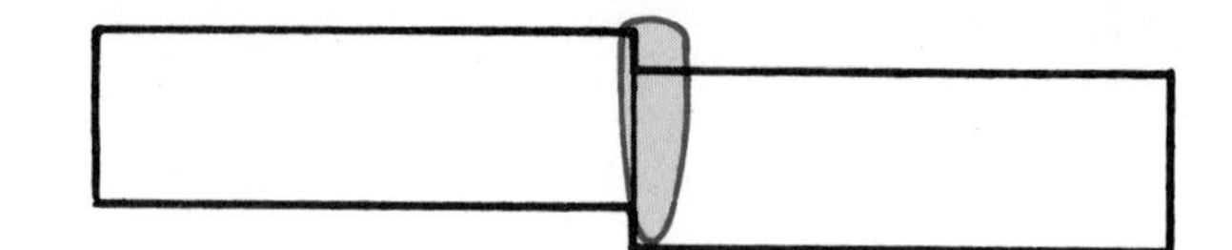

Fig. 6-23. Welds made on mismatched joints often fail below the rated load when placed into stressful conditions.

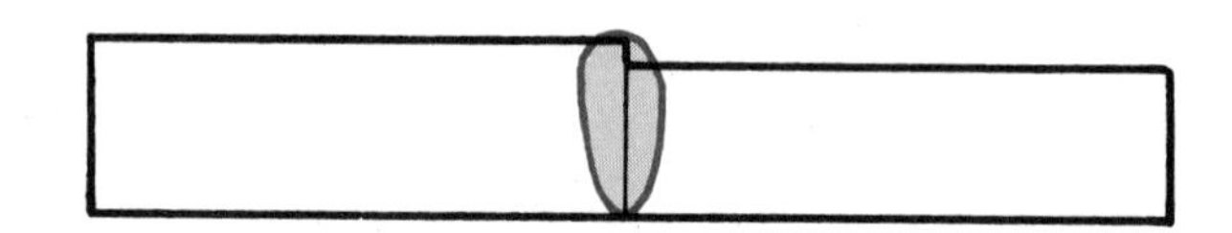

Fig. 6-24. Mating the joint at the bottom equalizes the load during stress.

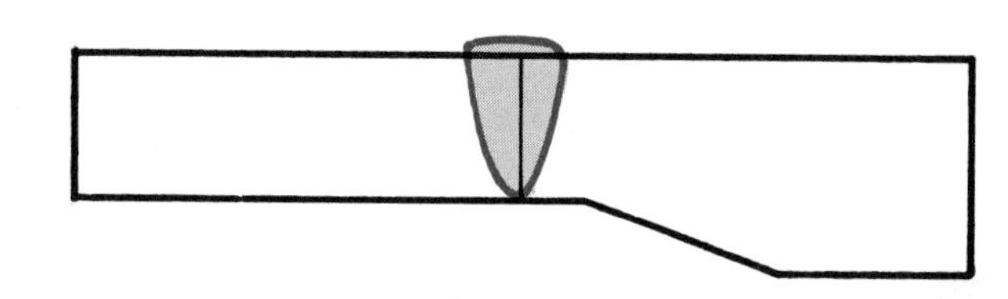

Fig. 6-25. Joints of unequal thickness absorb different amounts of heat and expand at different ratios. Equalize the heat flow by tapering the heavier material to the thickness of the thinner material.

Weld Shrinkage

Butt welds always shrink across (transverse) the joint during welding. For this reason, a shrinkage allowance must be made if the after welding overall dimensions have a small tolerance. Butt welds in pipe, tubing, and cylinders also shrink on the diameter of the material. This shrinkage is shown in Fig. 6-26. In areas where these dimensions must be held, a shrinkage test must be made to develop the amount of shrinkage. Fig. 6-27 shows how the test is made. Heavier materials will shrink more than thinner materials. Double groove welds will shrink less than single groove welds. This is because less welding is involved and less filler material is used.

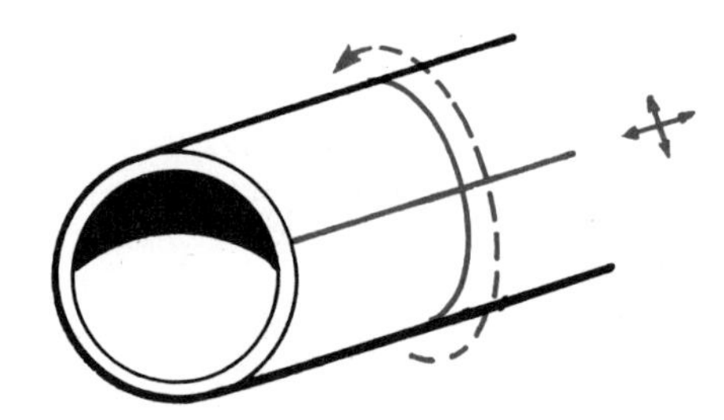

Fig. 6-26. Butt welds shrink during welding in both transverse and longitudinal directions.

Lap Joints and Welds

Lap joints may be either single, double fillet, plug slot, or spot welded. They require very little joint preparation. They are generally used in static load applications or in the repair of unibody automobiles. Where corrosive liquids are used, both edges of the joint must be welded, like in Fig. 6-28. One of the major problems with lap joint design is shown in Fig. 6-29 where the component parts are not in close contact. A bridging fillet weld must then be made which leads to incomplete fusion at the root of the weld and oversize fillet weld dimensions. When using this type of design in sheet or plate material, adequate clamps or tooling must be used to maintain contact of the material at the weld joint.

An interference fit eliminates this problem in assembly of cylindrical parts, Fig. 6-30. The inside diameter of the outer part is several thousandths of an inch smaller than the outside diameter of the inner part. Before assembly, the outer part is heated until it expands enough to slide over the inner part. As it cools it locks onto the smaller part.

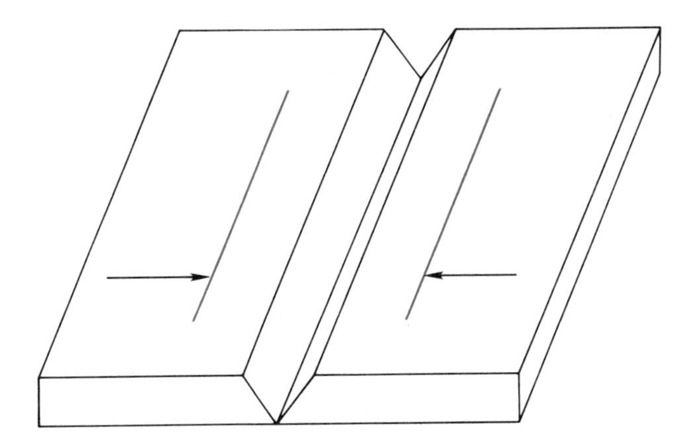

Fig. 6-27. Weld joint shrinkage can be determined in four steps. 1. Tackweld test joint together. 2. Scribe parallel lines, as shown, with approximately 2 in. centers. Record this dimension. 3. Weld joint with test weld procedure. 4. Measure linear distance and compare with original dimension.

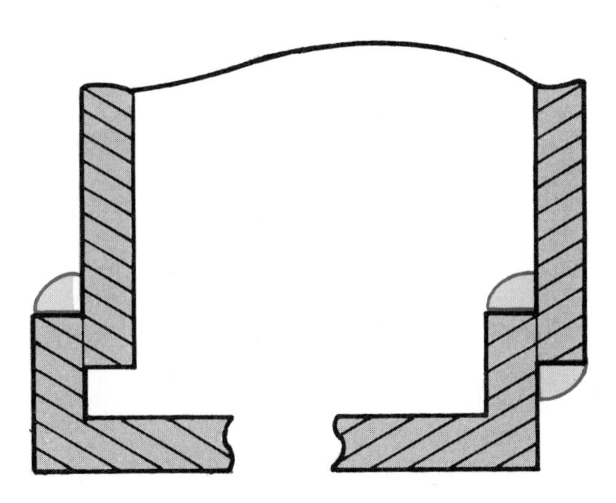

Fig. 6-28. Corrosive liquids must not be allowed to enter the penetration side of the weld joint.

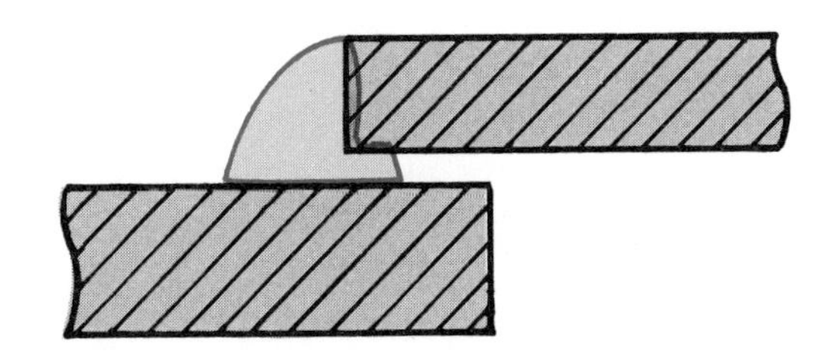

Fig. 6-29. Lap joint problem areas as a result of improper fitup.

T Joints and Welds

T Joint designs are used for joining parts at angles to each other. Depending on the use of the joint, they may be made with a single fillet, double fillet, or a groove and fillet weld combination. Fig. 6-31 shows how these designs may be used.

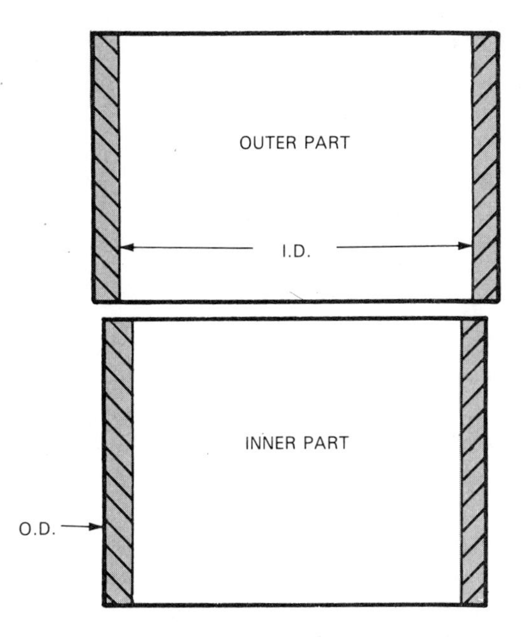

Fig. 6-30. The diameters of the component parts to be assembled may be determined by using a "Pi" tape around the inner and outer cylinder components. The tape measures in thousandths of an inch and full inches.

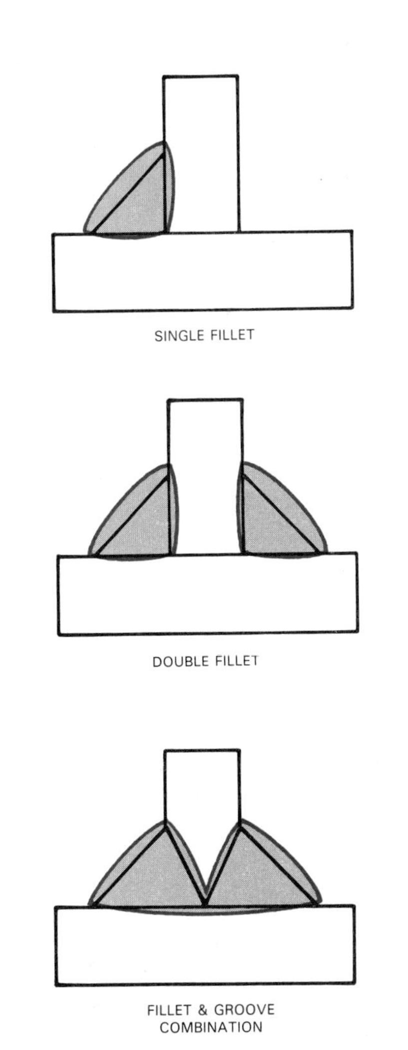

Fig. 6-31. Various types of T joints and welds.

Fillet welds are made to specific sizes determined by the allowable design load. They are measured as shown in Fig. 6-32. Where design loads are not known, the "rule of thumb" may be used for determining the fillet size. In these cases, the fillet weld leg lengths must equal the thickness of the thinner material.

The main problem in making fillet welds is the lack of penetration at the joint intersection. To prevent this condition, always make stringer beads at the intersection. Weave beads are prone to lack penetration on fillet welds.

Corner Joints and Welds

Corner joints are similar to T joints as they consist of sheets or plates mating at an angle to one another. They are usually used in conjunction with groove welds, and fillet welds. Many different designs may be used, some of which are shown in Fig. 6-33. When using thinner gauges, assembly of component parts may be difficult without proper tooling. Tackwelding and welding often cause distortion and buckling of thinner materials. For the most part, this type of joint design should be limited to heavier materials in structural assemblies.

Edge Joint and Welds

Edge welds are used where edges of two sheets or plates are adjacent and in approximately parallel planes at the point of welding. Fig. 6-34 shows several types of edge weld designs. These designs are common only in structural use. Since the weld does not penetrate completely through the joint thickness, it should not be used in stress or pressure applications.

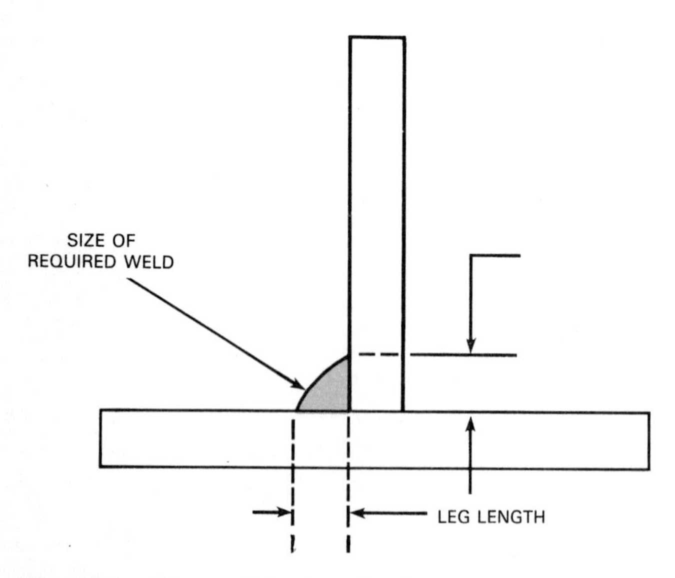

Fig. 6-32. Fillet weld leg lengths should be equal distance from the root of the joint. Unequal leg length, unless otherwise specified, will not carry the designed load and may fail when stressed.

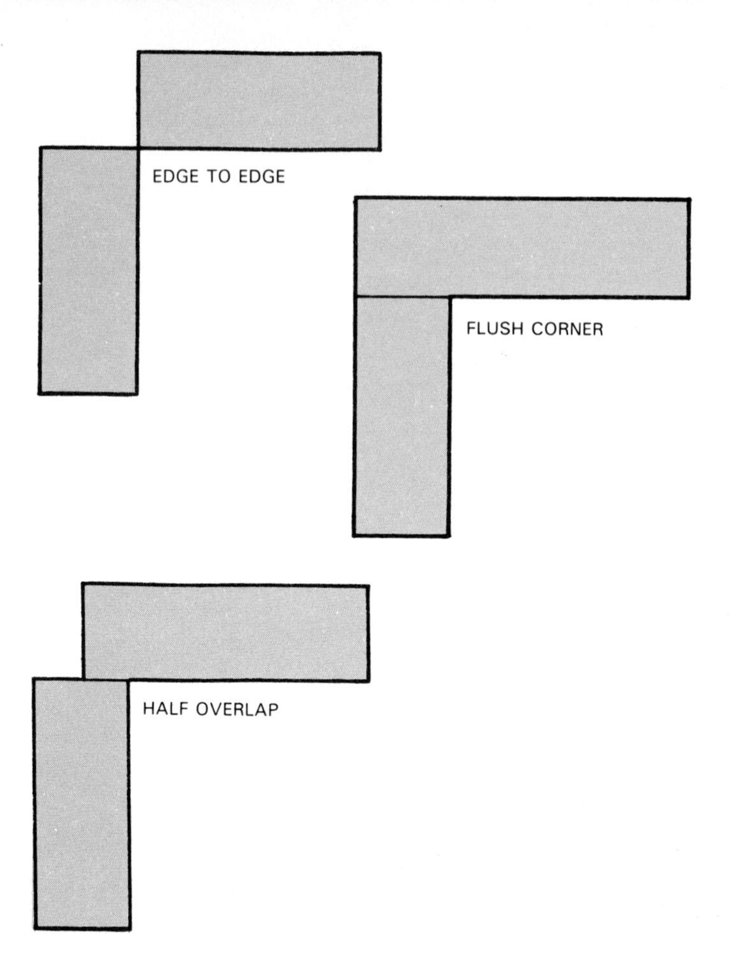

Fig. 6-33. Common corner weld joint designs which may be used in fabrication of component parts.

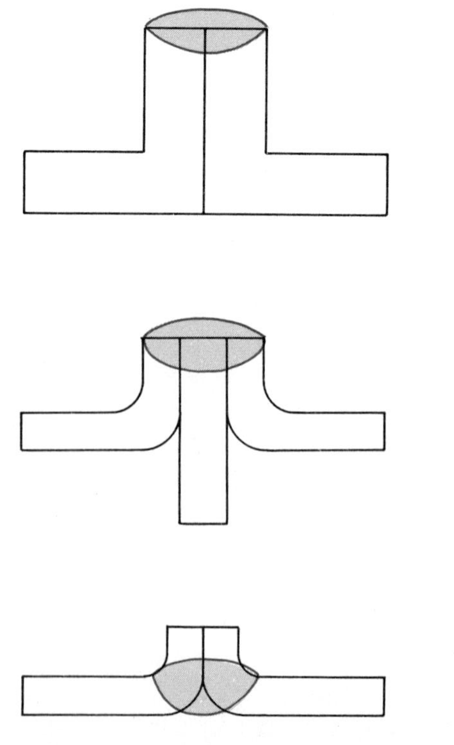

Fig. 6-34. Common edge weld joint designs which may be used in fabrication of component parts.

Special Designs and Procedures

Special designs are often used in the fabrication of a weldment where:

1. Welds cannot be thermally treated after welding because of configuration or size. A typical joint design, pictured in Fig. 6-35, is used to achieve full mechanical values required in the weld.
2. Joining of dissimilar materials can be done by buttering the face of one material to match the other, as shown in Fig. 6-36.
3. Special procedures and tooling may be used to provide a preheating, interpass, and postheating operation to control grain size. Preheating is generally used to slow down the cooling rate of the weld to prevent cracking. Interpass temperature is the minimum temperature at which a weld can be made on a multi-pass weld. Individual chapters regarding the welding of various metals will define the requirements for preheating, interpass, and postheat temperatures. These temperatures may be checked by the use of special crayons, paints, or pellets like those in Fig. 6-37.
4. Special procedures, tooling, and chill bars may be used to localize and remove welding heat during the welding application. Fig. 6-38 shows an application of tooling used to remove heat from the part.

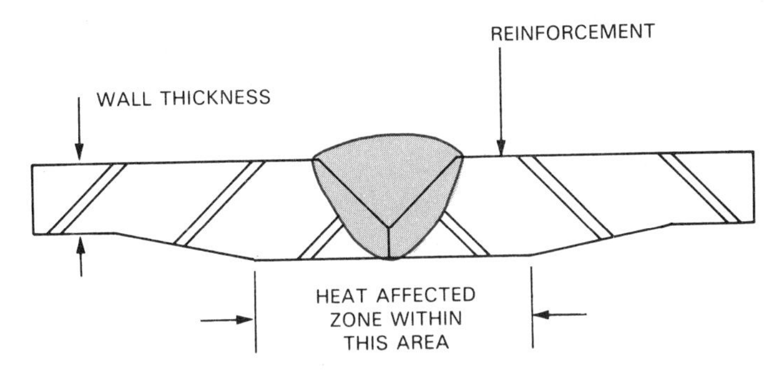

Fig. 6-35. The joint thickness and the filler material tensile strength is equivalent to the strength of the base material in this design.

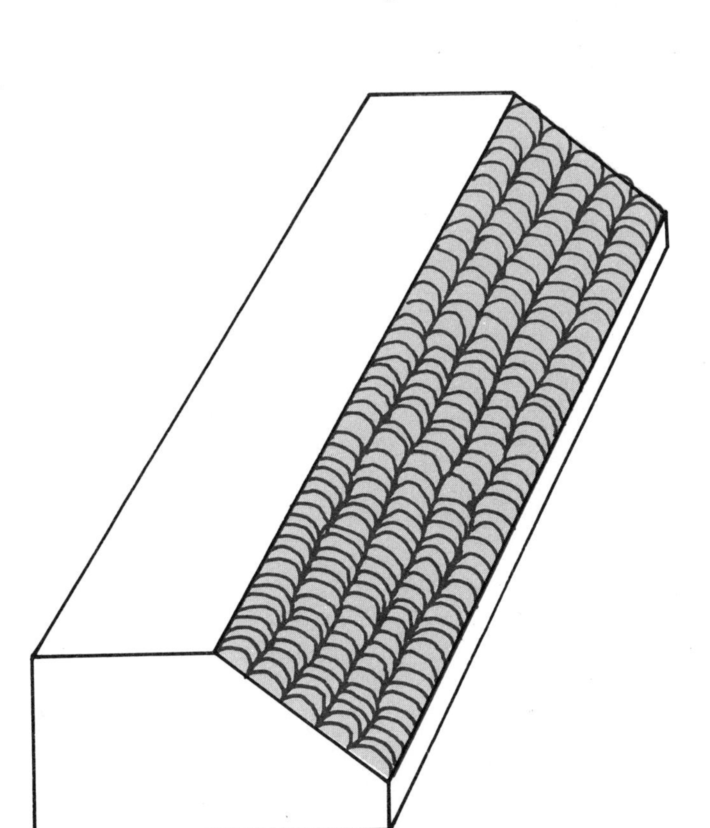

Fig. 6-36. Buttering techniques are commonly used to adapt dissimilar metals for welding.

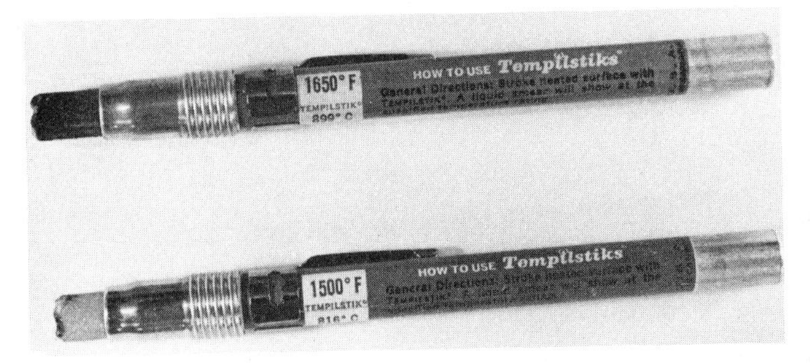

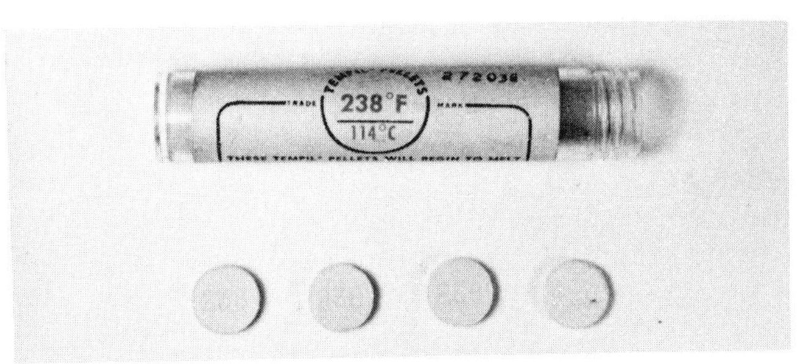

Fig. 6-37. Temperature crayons and pellets may be used to determine preheating temperatures prior to welding.

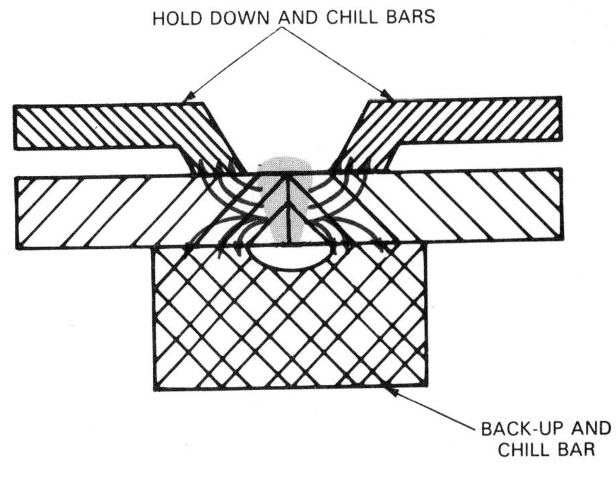

Fig. 6-38. Tooling and bars used to remove heat from the weld and resist heat flow into the base material are called CHILL BARS.

REVIEW QUESTIONS—Chapter 6

1. What are the five basic types of joints?
2. The basic types of joints are used for all welding processes for ________ types of welds.
3. Welds made from both sides of the joint are called ________ welds.
4. ________ welds are often used where dissimilar materials are joined.
5. Weld material placed on the surface of a joint for protection of the base material is called ________ or ________.
6. The oxyacetylene and the plasma arc processes are ________ cutting processes which leave an oxide film on the surface of the completed cut.
7. Where high strength welds are required, a ________ joint weld is used with 100 percent penetration.
8. Unless tackwelds are made properly, they often ________ during the welding operation.
9. Fillet welds are always measured by the length of the ________.
10. Where fillet weld dimensions are not specified, the size of the weld should equal the thickness of the ________ material.
11. The condition of a butt joint where neither the top nor the bottom edge of the material is flush is called ________.
12. ________ joints always shrink across the weld joint.
13. The main problem in welding fillet welds is the ________ ________ ________ into the root of the joint.
14. Edge type joints without full penetration of the weld should not be used in ________ ________.
15. ________ bars are used during welding to localize and remove heat from the weld area.
16. The American Welding Society defines the different positions of welding. The four positions include:
    A. ________ position.
    B. ________ position.
    C. ________ position.
    D. ________ position.

Chapter 7

WELDING PROCEDURES AND TECHNIQUES

After reading this chapter, you will be able to:

- Adjust welding current and voltage parameters to produce desired qualities in welds.
- Vary electrical stickout, travel speed, direction of travel, torch angle, and bead pattern for desired effects in weld quality.
- Determine values for a welding schedule based on results from test welds.

BASIC OPERATION

The GMAW process is considered either a semiautomatic or a fully automatic operation. In the fully automatic operation, all of the parameters and variables are controlled by the machine and operator skill is not essential. However, in the semiautomatic operation, the hands-on skill of the welder becomes important in the final quality of the weld. The welder must be able to set up the machine properly. Then, the welder must operate the torch in such a manner that the weld will meet fabrication requirements.

To accomplish this task, practice combined with the ability to change machine parameters and welding techniques is needed so the desired weld can be produced.

Welding Parameter and Variables

The major parameters are established after the type and size of filler metal and the mode of welding have been established. These parameters include:

1. Welding current/amperage (wire speed). The welding current is established by the wire speed with a constant voltage machine. Increasing or decreasing the wire speed will result in an increase or decrease in welding current.

   Wire speed is always established in inches per minute (IPM). This parameter may be determined by using the following sequence:

   a. Turn machine equipment on.
   b. Depress arc start switch (this starts wire feed) and run for 10 seconds.
   c. Measure wire length from end of contact tip.
   d. Multiply wire length by six (10 seconds = 1/6 of a minute).

   REMEMBER: Keep the end of the wire away from the ground. If contact is made, an arc will occur.
2. Welding voltage. The welding voltage is the actual arc gap established between the end of the electrode and the workpiece. This parameter is set on the welding machine. The machine then automatically changes the amperage output to maintain this preset arc voltage.

Reference charts for the initial welding parameters for steel, stainless steel, and aluminum are located in the reference section.

The major variables that are to be established include:

1. Electrical (or weld wire) stickout. Electrical stickout is the distance from the contact tip to the end of the electrode. See Fig. 7-1. The welder controls this

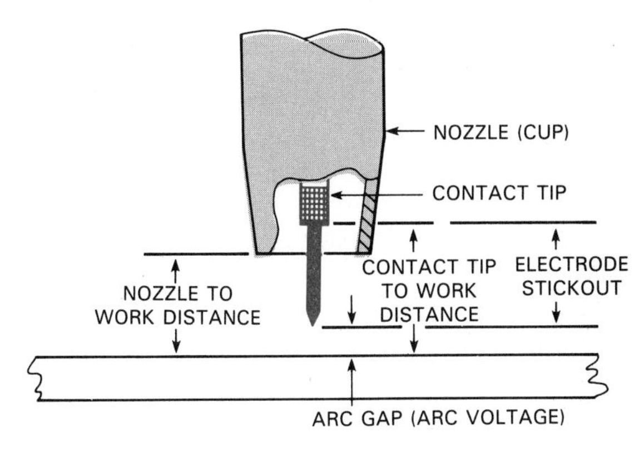

Fig. 7-1. Electrical stickout. This dimension is the actual electrode extension from the end of the contact tip to the end of the wire.

variable through the handling of the torch while welding. Changing this extension from the original established dimension has a marked effect on the melting of the electrode. Increasing the stickout will cause more resistance heating of the wire and a lower welding current flow. This is useful when welding on gapped joints to prevent burnthrough. Conversely, decreasing the stickout makes less wire available for resistance heating. This increases current flow for greater penetration. Fig. 7-2 shows this relationship.

2. Travel speed. The travel speed of the welding torch must be regulated to produce a good weld. In machine welding, the speed is set in inches per minute (IPM). In manual welding, the welder controls this variable. The welder must be able to adapt to changing shapes, improper fitup, gaps, and other variables as the weld continues. This is where welder skill is very important to the quality of the weld.
3. Direction of travel. The GMAW operation can be done with the welding torch pointing back at the weld and the weld progressing in the opposite direction. This is called BACKHAND or PULL welding and is shown in Fig. 7-3. Some of the results of this method include:
   A. More stable arc.
   B. Less spatter.
   C. Deeper penetration.

   The other method of torch manipulation is to point the torch forward in the direction of travel. This method is called FOREHAND or PUSH welding. This is shown in Fig. 7-4. Some of the results of this method include:
   A. More spatter.
   B. Less penetration.
   C. Better welder visibility of the weld seam.
   D. Good cleaning action of the arc when welding aluminum.
4. Torch angle. Torch angles are defined in either longitudinal (along the weld) or transverse (across the weld) angle dimensions. These are found in Fig. 7-5. Torch angle affects weld penetration, bead form, and final weld bead appearance. The angles shown for the various welds are to be used as a starting point for your weld. As the weld progresses, changes may be required in the various torch angles to maintain good puddle control and weld bead shape.

   Fig. 7-6 shows the torch angles and bead placement for flat groove welds. Fig. 7-7 shows the torch angles and bead placement for horizontal groove welds.

   Fig. 7-8 shows the torch angles and bead placement for horizontal fillet welds. Fig. 7-9 shows the torch angle for groove and fillet welding vertical up.

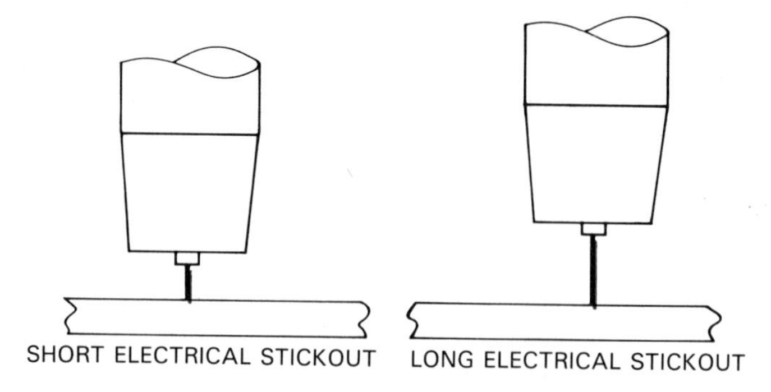

Fig. 7-2. Electrical stickout variations. Use these variations from the welding procedure dimensions to increase or decrease welding heat into the molten pool.

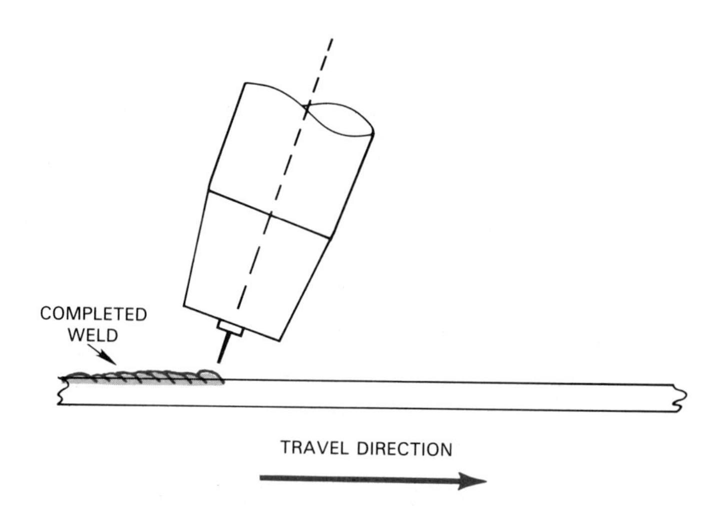

Fig. 7-3. Backhand or pull type welding technique. This technique requires more skill on the part of the welder as the weld joint is difficult to see because the gas nozzle is between the welder and the joint.

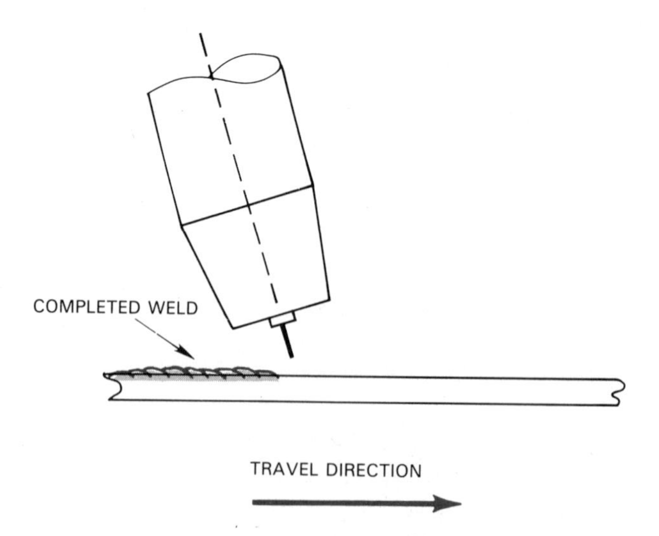

Fig. 7-4. Forehand or push type welding technique. The weld joint is directly in front of the weld wire, therefore, this technique is easy to use. When using this technique, remember that there is less penetration into the joint and spatter will increase.

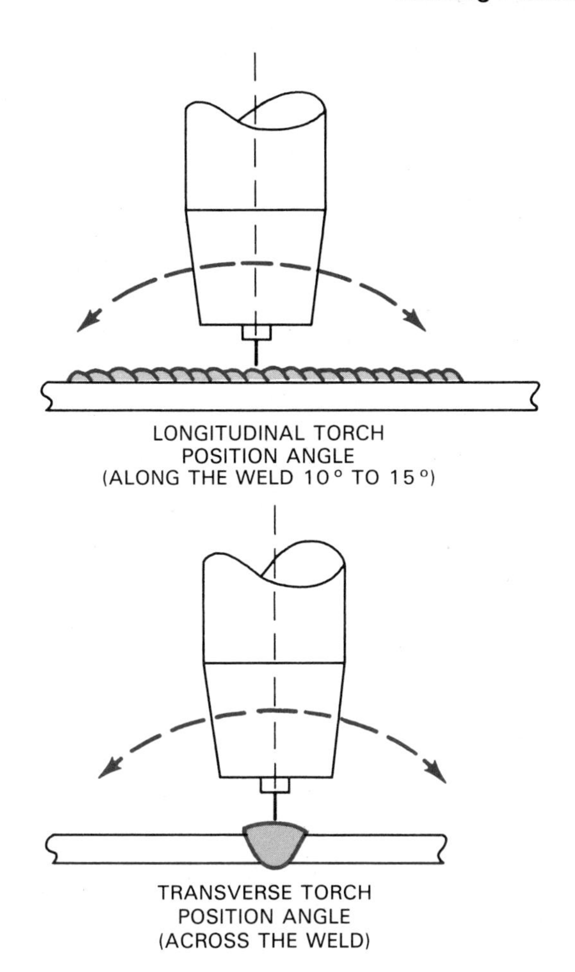

Fig. 7-5. Torch angles. Torch angles along the joint are always called longitudinal angles. Torch angles across the joint are always called transverse angles.

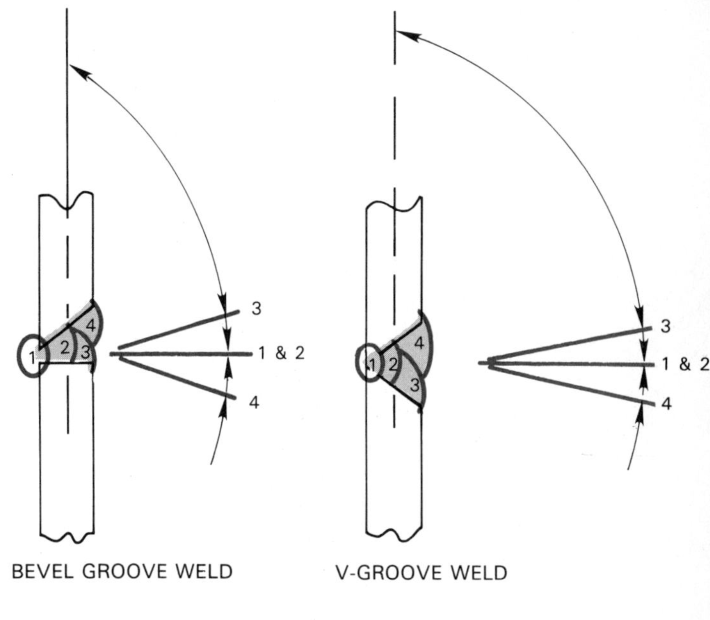

Fig. 7-7. Horizontal groove torch angles and bead placement. Longitudinal angles are 10-15° from the vertical position. The joint on the left is a bevel groove joint with the torch transverse angles shown for initial setup only. Remember to make the weld beads smaller when welding out of position and always make stringer beads. (Do not make wash beads.) Fill each layer from the bottom upwards. The joint on the right is a V groove joint with the torch transverse angles shown for initial setup only. The same rules apply as in a bevel groove joint.

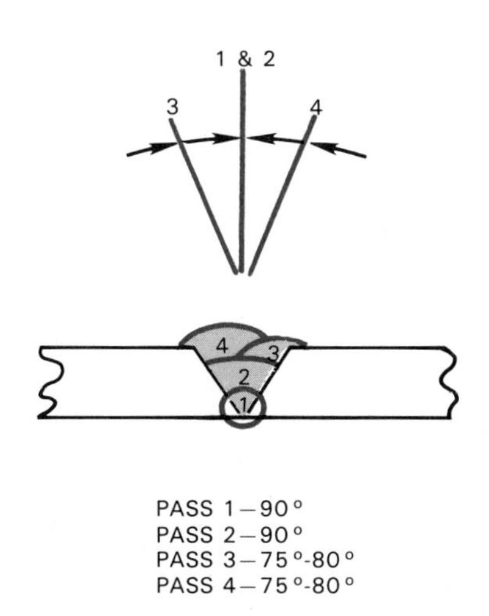

Fig. 7-6. Flat groove weld torch angles and bead placement. Longitudinal angles are 10-15° from the vertical position. The angles shown are transverse angles and are only for initial welding setup. Each variation in groove angle or thickness dimension will affect the required torch angle. Remember, the molten weld pool must flow into the previous pass and the side walls of the joint. Develop your welding skill so that you may be able to properly fill the groove joint with a sound weld.

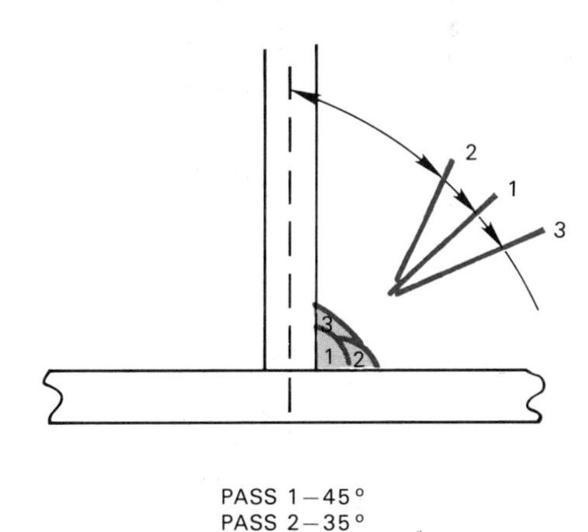

Fig. 7-8. Horizontal fillet weld torch angles and bead placement. Longitudinal angles are 10-15° from the vertical position. The angles shown are transverse angles and are only for initial welding setup. For a single pass weld, use the 45° angle and use the other angles for a multiple pass weld. On a multiple pass weld, always make the second pass on the bottom. The completed weld should have: 1—Slightly convex crown. 2—Equal length legs. 3—The size of the fillet should be equal to the thickness of the thinnest material used in the joint.

Fig. 7-9. Vertical groove and fillet weld torch placement. Welding vertical up. The groove weld is shown on the left. The fillet weld is shown on the right. The angles shown are for longitudinal torch placement and are for initial welding setup only. The bead placement for the fillet weld is the same as for flat and horizontal welds. Wash beads may be used after the root pass is made and the joint access is sufficient to accept the additional metal.

Fig. 7-10. Vertical groove and fillet weld torch placement. Welding vertical down. The groove weld is shown on the left. The fillet weld is shown on the right. The angles shown are for longitudinal torch placement and are for initial welding setup only. The bead placement for fillet welds is the same as for flat and horizontal welds. Make stringer beads and keep the wire on the front edge of the molten pool.

Fig. 7-10 shows the torch angle for groove and fillet welding vertical down.

5. Weld bead patterns. Two types of patterns are used for depositing metal. They include:
   A. Stringer bead pattern. In this pattern, travel is along the joint with very little side-to-side motion. It may be made with a small zig-zag motion or in a small circular motion. These motions are pictures in Fig. 7-11.
   B. Weave bead pattern. This may also be called a wash bead pattern or an oscillation bead pattern. The weld is wider than a stringer bead and requires a dwell (wait) of the torch at the end of each weave pattern to fill the weld metal into the weld without undercut. This is shown in Fig. 7-12.

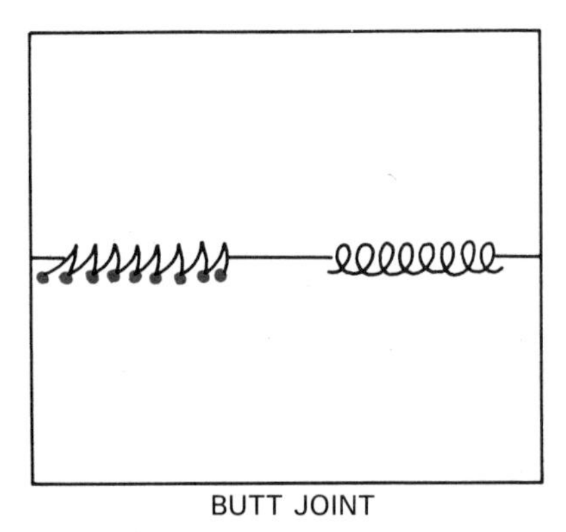

Fig. 7-11. Stringer weld bead patterns. The bead on the left is called a backstep or zig-zag technique and requires a dwell point on each swing backward. These points are marked with a colored dot. The circular bead technique shown on the right does not have any dwell points. As you practice these various techniques, vary your forward speed to maintain an even bead contour and shape.

Welding Schedules

The actual welding of component parts requires the combination of many parameters and variables. These many areas cannot be remembered for all the many materials, wire sizes, types of gases, and techniques used to make the weld. Therefore, a listing of all of these areas can be placed on a form called a weld schedule, Fig. 7-13. To determine the actual values placed on the form, select an initial setting from the charts shown in the reference section for the type and thickness of material involved. The next step is to clean the material and make test welds. During the test weld period, adjust the machine and weld techniques to produce the desired weld. Fig. 7-14 shows the various types

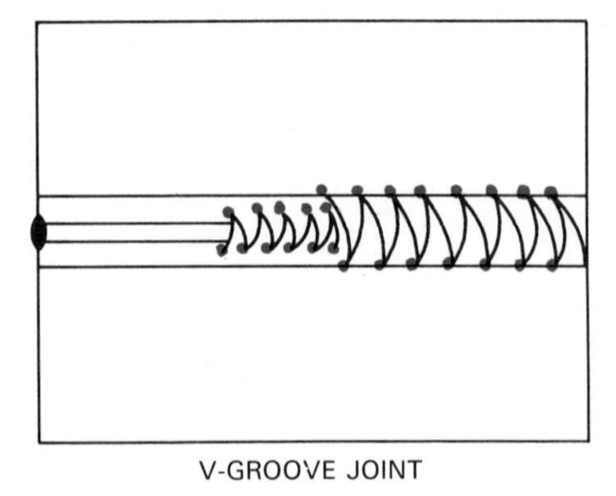

Fig. 7-12. Weave bead patterns. The weave bead pattern is prone to undercut at the outer edges of the weld and requires a hold or dwell point as indicated by the colored dot.

GAS METAL ARC WELDING SCHEDULE

Process Mode ________

Weld Type ________ Position ________

Base Metal Type ________ Base Metal Thickness ________

Wire Type ________ Wire Dia. ________ Wire Speed (IPM)________

Gas Type ________% Gas Type ________% Flow Rate (CFH) ________

Nozzle Type________ Nozzle Dia. ________

Contact Tip Type (long) ____ (short) ____ (standard) ____ Dia. ____

Amperes ________ Volts (arc) ________ Stickout________

Spotweld Time (seconds) ________ Burnback Time (seconds) ________

Travel (forehand) ____ (backhand) ____(vertical up)____ (down)____

Pre-weld Cleaning Type________________

Fig. 7-13. Weld schedule. Forms such as this are used by industry to record data for ready reference for each job they may do.

WELDING VARIABLES TO CHANGE	DESIRED CHANGES							
	PENETRATION		DEPOSITION RATE		BEAD SIZE		BEAD WIDTH	
	↑	↓	↑	↓	↑	↓	↑	↓
CURRENT & WIRE FEED SPEED	↑	↓	↑	↓	↑	↓	*	*
VOLTAGE	+	+	*	*	*	*	↑	↓
TRAVEL SPEED	+	+	*	*	↓	↑	↑	↓
STICKOUT	↓	↑	↑	↓	↑	↓	↓	↑
WIRE DIAM.	↓	↑	↓	↑	*	*	*	*
SHIELD GAS % CO_2	↑	↓	*	*	*	*	↑	↓
TORCH ANGLE	BACK HAND TO 25°	FORE HAND	*	*	*	*	BACK HAND	FORE HAND

* NO EFFECT
+ LITTLE EFFECT
↑ INCREASE
↓ DECREASE

Fig. 7-14. Adjustments in welding parameters and techniques. (L-Tec Co.)

of adjustments that can be made in procedures and techniques. Chapter 11 of this text shows many of the common defects that occur during welding and steps that can be taken to correct these conditions.

Each welder will have his or her own areas of technique. Therefore, do not expect each individual weld to be exactly like another. Machine welds can be duplicated, and the tolerance for the variables can be smaller. In the case of manual welds, you should allow a larger tolerance in variables and procedures.

REVIEW QUESTIONS—Chapter 7

1. The GMAW process is considered either a semiautomatic or ________ automatic welding operation.
2. The hands on skill of the welder is very important when welding in the ________ mode.
3. The two major parameters that must be established on the equipment prior to welding are ________ ________ and ________ ________.
4. Electrical stickout is the length of wire that extends out of the contact tip. Increasing this length dimension ________ the amount of current flow making the molten pool become sluggish.
5. Using the ________ technique for welding decreases penetration, increases spatter, and is always used for welding aluminum.
6. Using the ________ technique for welding increases penetration, and has minimum spatter.
7. Torch angles used in GMAW are defined as along the weld or ________ angles.
8. Torch angles used in GMAW are defined as across the weld or ________ angles.
9. When welding on the joint without any or minimal side-to-side motion, this type of weld bead pattern is commonly called a ________ ________.
10. When welding on the joint with side-to-side motion, this type of weld bead pattern is commonly called a ________ ________. This type of pattern may also be called a ________ ________.
11. Information regarding the weld joint design and the welding parameters and variables may be placed on a form called a ________ ________.

Chapter 8

WELDING THE CARBON STEELS

After studying this chapter, you will be able to:

- Use the appropriate filler metal for different conditions.
- Prepare a joint for welding.
- Apply preheat, interpass temperature, and postheat in correct proportions on various joints.
- Follow procedures for tooling applications.
- Employ welding procedures in sequence.
- Spot weld carbon steel.
- Plug and slot weld carbon steel.

BASE MATERIALS

Many types and grades of steel are included in the basic steel family. Carbon steels are magnetic and melt at approximately 2500 °F. They are identified as the group of steels which contain:

1. Carbon—1.70 percent max.
2. Manganese—1.65 percent max.
3. Silicon—0.60 percent max.

Carbon steels may be further classified as:

1. Low carbon steel—up to 0.14 percent carbon.
2. Mild carbon steel—0.15 to 0.29 percent carbon.
3. Medium carbon steel—0.30 to 0.59 percent carbon.
4. High carbon steel—0.60 to 1.70 percent carbon.

Low Alloy Steels

Alloy steels contain varying amounts of carbon in addition to many alloying elements. These elements include chromium, molybdenum, nickel, vanadium, and manganese. The elements increase the strength and toughness of the material. In some cases, they also increase corrosion resistance.

Quenched and Tempered Steels

Quenched and tempered steels and high strength steels (HSS) are produced by a number of steel companies. They have many trade names. These steels are hardened and tempered for specific mechanical values at the mill. By using these processes, strength, impact resistance, corrosion resistance, and other properties can be improved. Manufacturers of these metals should be consulted prior to welding. This will assure proper welding so special features can be maintained.

Chrome Moly Steels

The alloys known as the chromium molybdenum steels are used in applications requiring high strength. These steels may be annealed, hardened, or tempered.

FILLER MATERIALS

Most common filler materials used for welding mild steels are defined in American Welding Specification A5.18. A code is used to identify different filler materials. A typical example is E70S-1:

The letter "E" stands for electrode.

The number "70" means the wire has 70,000 pounds tensile strength.

The letter "S" labels the wire as a solid wire.

The number "1" identifies the class of wire. Each class is defined by the properties a wire will produce when welded with a specific shielding gas. The chemical compositions of various steel welding wires are shown in the reference section and include:

1. E70S-2. This wire is heavily deoxidized, meaning that it contains agents to fight oxidation. It is designed for producing sound welds in all grades of carbon steel: killed, semikilled, and rimmed. Due to added deoxidants (aluminum, zirconium, and titanium), it can weld carbon steel with a rusty surface. Argon-oxygen, argon-carbon dioxide, and carbon dioxide shielding gases can be used. In general, an extremely viscous (not fluid) weld puddle is produced. This makes it an ideal wire for short-arc welding out of position. A high oxygen or carbon dioxide content helps improve the wetting action of the puddle.

2. E70S-3. This wire is one of the most widely used wires with GMAW. E70S-3 wires can be used with either carbon dioxide, argon-oxygen, or argon-carbon dioxide in killed and semikilled steels. Rimmed steels should be welded with only argon-oxygen or argon-carbon dioxide. High welding currents used with carbon dioxide shielding gas may result in low strength. Either single pass or multipass welds can be made with this electrode wire. The tensile strength for a single pass weld in thin gauge low and medium carbon steels exceeds the base material while ductility is adequate.

   In a multipass weld, the tensile strength will range between 65,000 and 85,000 psi depending on the base metal dilution and type of shielding gas. The weld puddle is more fluid than that of the E70S-2. The E70S-3 has better wetting action and flatter beads. This wire has its greatest application on automobiles, farm equipment, and home appliances.
3. E70S-4. Wire electrode of this classification contains a higher level of silicon than the E70S-3. This improves the soundness on semikilled steels and increases weld metal strength. It performs well with argon-oxygen, argon-carbon dioxide, and carbon dioxide shielding gases. It also can be used with either spray or short-arc techniques. Structural steels such as A7, A36, common ship steels, piping, pressure vessel steels, and A515 Grades 55 to 70 are usually welded with this wire. Under the same conditions, weld beads are generally flatter and wider than those made with the E70S-2 or the E70S-3.
4. E70S-5. In addition to silicon and manganese, these wires contain aluminum (Al) as a deoxidizer. Because of high Al content, they can be used for welding killed, semikilled, and rimmed steel with carbon dioxide shielding gas at high welding currents. Argon-oxygen and argon-carbon dioxide may also be used. Short-circuiting type transfer should be avoided because of extreme puddle viscosity. Rusty surfaces can be welded using this wire with little sacrifice in weld quality. Welding is restricted to the flat position.
5. E70S-6. Of all the electrodes, this one is high in manganese content and has the most silicon. These elements serve as deoxidizers. Like the E70S-5, it yields quality welds on most carbon steels. This assumes the use of carbon dioxide shielding gas and high welding currents. Also used are argon-oxygen mixtures containing five percent or more oxygen on high speed welding. Because this wire contains no aluminum, the short-arc technique is possible using carbon dioxide or argon-carbon dioxide shielding gases. Like the E70S-4, the weld puddle is quite fluid.
6. E70S-1B. This wire contains silicon and manganese as deoxidants plus molybdenum for increased strength. Welds can be made in all positions with argon-carbon dioxide and carbon dioxide shielding gases. Argon-oxygen is permitted for the flat position. Maximum mechanical properties are obtained with argon-oxygen and argon-carbon dioxide mixtures. Welding can be done over slightly rusted surfaces with some sacrifice in weld quality. This wire is mostly used for welding low alloy steels such as AISI 4130.

Joint Preparation and Cleaning

Joint edges prepared by the thermal cutting processes form a heavy oxide scale on the surface like in Fig. 8-1. This oxide should be completely removed prior to welding. This prevents porosity and dross within the weld. To a great extent, GMAW nullifies foreign material on the base metal. However, good welds cannot be made over oxide scale.

Remove this scale and rough edges with a grinder. Foreign material next to the weld joint is removed by sanding, Fig. 8-2. When making multiple passes, always clean the weld with a wire brush to remove oxides and foreign material. Where wire brushing does not remove

Fig. 8-1. The oxide scale and small gouge indentations were formed on this bevel cut by the oxyacetylene cutting process.

Fig. 8-2. The material has been sanded to "bright metal" to prepare the part for welding. This type of sander is often called a "PG" wheel.

the oxide scale, it must be removed by chipping or grinding. Failure to remove these oxides may result in lack of fusion when the next pass is made.

Many materials are received from the mill with a coating of oil to prevent rusting. When good quality welds are desired, remove the oil before welding. This oil is removed with cleaning solvent. This helps prevent porosity.

Preheating, Interpass Temperature, and Postheating

Underbead cracking or lack of toughness in the heat-affected zone is not usually encountered when welding mild steels containing less than 0.20 percent carbon and one percent manganese. These steels can be welded without preheat, postheat, or special welding procedures. This is true when the joint thickness is less than one inch and restraint is not severe.

Low alloy steels, like chromium-molybdenum steels, need to be preheated. Not enough preheating results in hard heat-affected zones after welding. These zones are caused by the base material cooling too rapidly. This type of cooling allows the formation of hard martensite grain structures. (Martensite is a type of grain structure which is very hard.) Preheating the base metal from 200 °F to 400 °F slows this cooling rate. This prevents the formation of martensite.

Carbon steels which contain over 0.30 percent carbon or have severe joint restraint must be preheated. They also require maintaining the interpass temperature. Finally, these steels must be postheated to prevent cracking. Steel manufacturers should be consulted for recommended temperatures.

Oxygen-acetylene torches are often used to supply preheat and interpass heat to the weld area. Temperature crayons, paints, or pellets are used to indicate the desired temperatures. Always use a neutral flame when preheating to prevent the torch from adding carbon or oxygen to the weld area.

Postheating the weld is a sequence of applying local heat to the welded area. The temperature is allowed to drop slowly, reducing the chances of cracking near the weld.

Stress relieving reduces weld stresses. This process requires that the weldment's temperature be maintained at approximately 1150 °F, one hour for each inch of thickness. The weldment is then allowed to gradually cool in still air.

Quenched and tempered steels must be welded with a procedure that prevents excessive heating of the material. Steel manufacturers have established maximum temperatures for these metals. These temperatures must be followed to achieve desired values.

Tooling

Tooling for GMAW is basically the same as for other processes with one exception. That is, in some welding modes considerable spatter is created. This means tooling near the weld must be made of copper, aluminum, or stainless steel to minimize spatter pickup. These materials also reduce the possibility of magnetic arc blow (the altering of the arc pattern due to magnetism).

The major items to be considered in tool design include:

1. Alignment of the components to be welded.
2. Heat control of the weld zone.
3. Positioning of the joint for welding.
4. Assembly (loading) and disassembly (unloading) of the components.
5. Providing atmosphere to prevent contamination of the weld crown and root.
6. Accessibility for the welding torch to the weld joint.

Welding Procedures

1. Select the filler wire and shielding gas to match the properties of the base metal.
2. Select welding parameters for the material thickness from setup charts in the reference section.
3. Establish the welding parameters on the power supply and wire feeder.
4. Set the shielding gas flow rate (CFH) and select a gas nozzle for the operation involved.
5. Select and install contact tip for process mode and diameter of wire.
6. Make a test weld on scrap material to determine that all of the required parameters and variables are correct.

SPOT WELDING THE CARBON STEELS

Spot welding can be used as a primary method of fusing components together. It may also be used to hold components together for the final welding operation. The various uses of the process are shown in Fig. 8-3. Special equipment required to accomplish the operation includes:

1. Welding timer. This controls the length of time the welding current is on.
2. Shielding gas post flow timer. This controls the length of time the gas will flow after the arc is stopped.

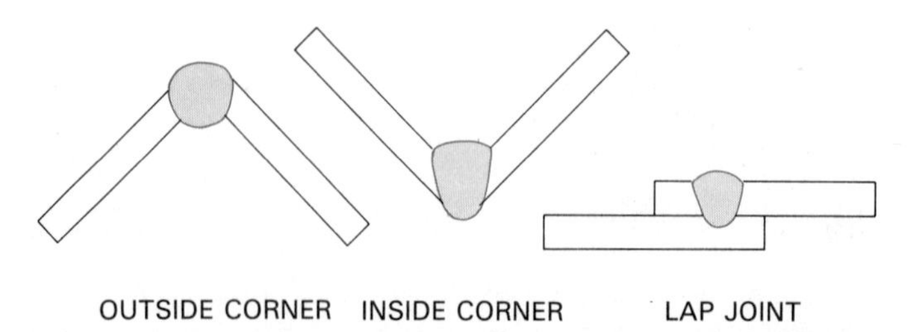

Fig. 8-3. Spot welding uses include lap and corner joints.

3. Anti-stick timer (commonly called a burn-back timer). This operates the electrical current flow after the wire stops. This allows the wire to continue melting after the wire feed stops, thus, preventing the freezing of the wire into the molten puddle.
4. Specially designed gas nozzles are used for the application of spot welding. The nozzles are designed to fit individual torches to maintain the contact tip to work distance. They have slots or gaps in the open end to allow the escape of shielding gases during the welding operation. Overlap, inside corner, and outside corner joint gas nozzles are shown in Figs. 8-4, 8-5, and 8-6, respectively.

Cleaning and Fitup

Lap joints MUST be cleaned on both pieces of metal where the spotweld will be located. This area should be prepared to bright metal before assembly. (Bright metal is metal that has been abrasively treated, thus, making the surface shiny.) The sander in Fig. 8-2 may be used in this part of the operation. Any oil on the surface is removed with solvent. The welding heat is transferred through the top plate into the bottom plate. Therefore, they must be clean and fit together tightly. Where spot welds are used in corner joints, the gap must be minimized or burnthrough will occur. A wide gap might also allow the wire to travel through the gap and become a whisker when the arc starts.

When making lap joints with different metal thicknesses, the thinnest piece must always be on the top side. If this cannot be done, it is suggested that a plug weld be made in place of the spot weld.

Welding Procedure

1. Select the wire to match the physical properties of the base metal. For the mild steels, use the E70S-4 or E70S-6 wires. They contain more deoxidizers to ensure sound welds without porosity.
2. Select the welding parameters for the thickness of the upper plate from the setup charts in the reference section.
3. Establish welding parameters on the power supply and wire feeder.

Fig. 8-4. The flat edge of the nozzle is held tight against the top plate during the welding operation. The hot gases escape through the slots.

Fig. 8-5. The inside corner nozzle is made to allow the gases to escape along these edges (arrows) of the nozzle tip.

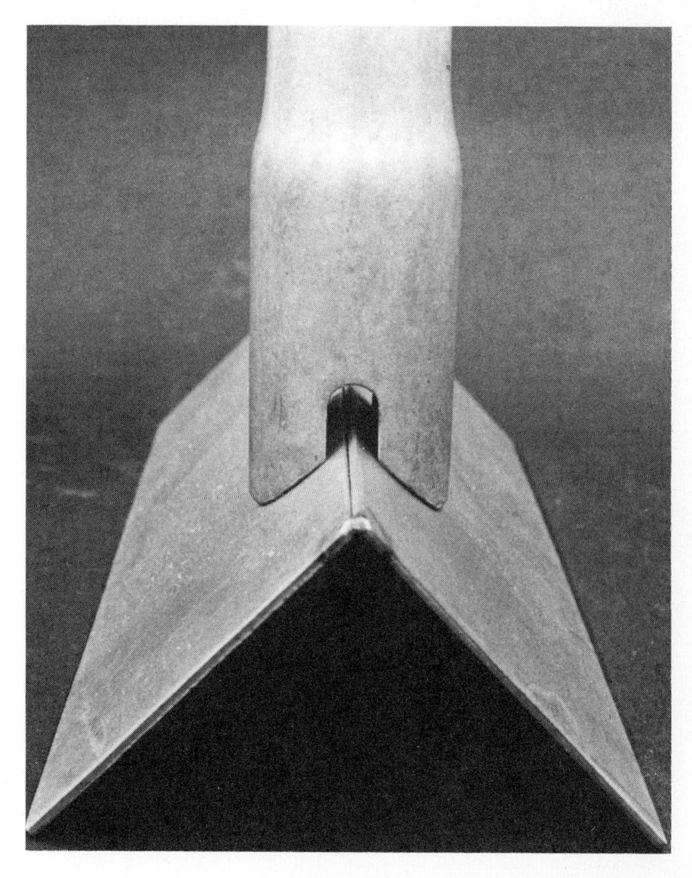

Fig. 8-6. Outside corner nozzle has slots for the escape of the hot gases.

4. Set shielding gas CFH flow rate and post flow time (if available).
5. Install contact tip according to the diameter of the wire. The contact tip must not extend to the end of the gas nozzle because clearance must be maintained for molten metal. The spot weld gas nozzle for the correct operation must also be installed.
6. Set burnback time so that the electrode does not melt back into the contact tip at the end of the weldment.
7. Using test material of the same thickness as the required weld, place the torch firmly against the joint seam and make a weld. The timer should be set for approximately one second. Inspect the completed weld for penetration of the nugget into the lower plate on a lap weld or penetration of the seam on a butt or corner weld. (On thinner materials, the spot weld may penetrate through the lower plate as shown in Fig. 11-25.) On heavier materials, a tensile test (a destructive test) of the weldment must be used to determine the actual shear strength of the nugget.
8. For more penetration of the nugget, increase wire feed, decrease welding voltage, or increase weld time. If penetration is too deep, reverse the operations outlined above.
9. Clean the gas nozzle often to remove weld spatter; use spatter protection spray on the inside of the gas nozzle.
10. Always trim the electrode before starting to weld so that the wire stickout does not extend to the edge of the gas nozzle.

Figs. 8-7, 8-8, and 8-9 show completed lap joint, inside corner, and outside corner spot welds, respectively.

Fig. 8-7. Completed lap joint spot weld.

PLUG AND SLOT WELDING THE CARBON STEELS

Plug and slot welds are often used in weldments where the material is too heavy for spot welds. They are also used where additional welding is required to reinforce the primary joint. Some typical uses of plug and slot welds are shown in Fig. 8-10.

These welds are made with the same procedures used for seam welding except the welder must manipulate the torch around the hole or slot. Thin gauge materials may be welded in one pass. Heavier materials, however, require several layers or passes. In these cases, always wire brush the completed weld before applying another pass or layer.

The major problem encountered in this type of weld is lack of fusion at the weld root. This is caused by improper torch angle as the welder tries to manipulate the torch around the hole. To minimize this condition, change gun angles as the weld progresses. Practice on sample weldments to develop the proper techniques. A typical plug weld is shown in Fig. 8-11.

Weld Defects

Do not feel frustrated when some of your welds contain defects. As your skills progress with practice, you

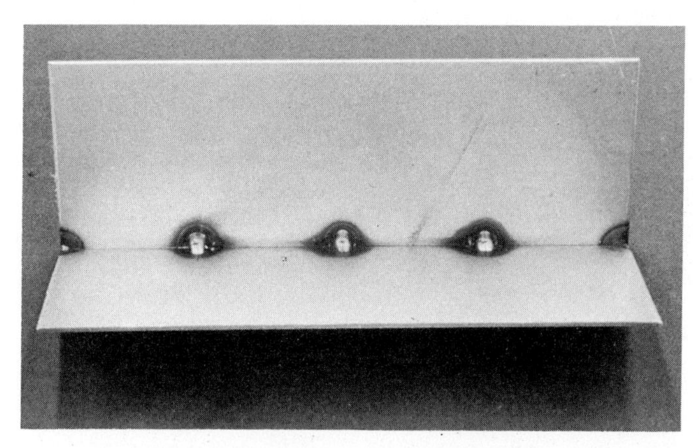

Fig. 8-8. Completed inside corner spot weld.

Fig. 8-9. Completed outside corner spot weld.

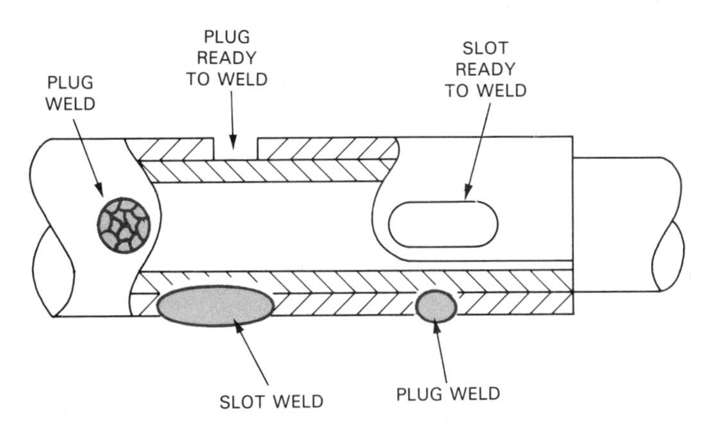

Fig. 8-10. Plug and slot welds may be made in many combinations.

Fig. 8-11. Completed plug weld.

will learn to avoid these mistakes. Chapter 11 discusses weld defects that are common when making various types of welds, regardless of the material. Refer to this chapter to learn the corrective action for the various defects you may encounter.

REVIEW QUESTIONS—Chapter 8

1. The carbon steels are __________ and melt at approximately 2500 °F.
2. Carbon steels have a maximum of 1.70 percent __________.
3. The quenched and tempered steels have been hardened and tempered to have specific __________ __________.
4. Most carbon steel electrodes are produced to an ________ ________ ________ specification.
5. What does each digit in E70S-2 stand for?
6. What complications can be caused by welding over oxides?
7. When welding multi-pass welds, always remove surface __________ before depositing additional passes or layers.
8. Coatings of ________ on steel to prevent rusting should be removed with solvent prior to welding.
9. Why should low alloy steels be preheated?
10. One of the major problems encountered in welding with steel tooling is the deflection of the arc by __________ __________ __________.
11. Lap joints must be cleaned before spot welding to __________ __________ prior to assembly and welding. When using various thicknesses of metal, the __________ material is always on top.
12. Spot welds on lap joints with thin materials may be __________ examined to determine weld penetration.
13. You have welded a spot weld and the penetration is not sufficient. To apply more heat to the weld, you should __________ the wire feed speed, decrease the __________, or adjust the weld timer for a __________ period of time.
14. The major problem encountered in slot welding is __________ __________ __________ at the root of the weld.
15. The welder needs considerable __________ to avoid lack of fusion when slot welding.

Chapter 9

WELDING STAINLESS STEELS

After studying this chapter, you will be able to:
- ■Select the proper filler wire for different types of stainless steel.
- ■Prepare a stainless steel joint for welding.
- ■Use appropriate tooling for stainless steels.
- ■Follow proper welding procedures for stainless steel.
- ■Spot weld stainless steel.
- ■Plug and slot weld stainless steel.

Today, there are many industrial applications for welding stainless steels. In order to meet vast demands, a variety of stainless steels have evolved. This section, however, will only discuss the chromium nickel group. When welding other stainless steels with GMAW, contact the manufacturer for recommendations or exact guidelines.

Base Materials

The base materials of the chromium nickel group are listed in Fig. 9-1. The main elements of this material are chromium and nickel which help the material resist

COMMERCIALLY WROUGHT STAINLESS STEEL IDENTIFICATION (AISI)

	COMPOSITION, PERCENT[a]							
Type	**C**	**Mn**	**Si**	**Cr**	**Ni**	**P**	**S**	**Others**
201	0.15	5.5-7.5	1.00	16.0-18.0	3.5-5.5	0.06	0.03	0.25 N
202	0.15	7.5-10.0	1.00	17.0-19.0	4.0-6.0	0.06	0.03	0.25 N
301	0.15	2.00	1.00	16.0-18.0	6.0-8.0	0.045	0.03	
302	0.15	2.00	1.00	17.0-19.0	8.0-10.0	0.045	0.03	
302B	0.15	2.00	2.0-3.0	17.0-19.0	8.0-10.0	0.045	0.03	
303	0.15	2.00	1.00	17.0-19.0	8.0-10.0	0.20	0.15 min	0-0.6 Mo
303Se	0.15	2.00	1.00	17.0-19.0	8.0-10.0	0.20	0.06	0.15 Se min
304	0.08	2.00	1.00	18.0-20.0	8.0-10.5	0.045	0.03	
304L	0.03	2.00	1.00	18.0-20.0	8.0-12.0	0.045	0.03	
305	0.12	2.00	1.00	17.0-19.0	10.5-13.0	0.045	0.03	
308	0.08	2.00	1.00	19.0-21.0	10.0-12.0	0.045	0.03	
309	0.20	2.00	1.00	22.0-24.0	12.0-15.0	0.045	0.03	
309S	0.08	2.00	1.00	22.0-24.0	12.0-15.0	0.045	0.03	
310	0.25	2.00	1.50	24.09-26.0	19.0-22.0	0.045	0.03	
310S	0.08	2.00	1.50	24.0-26.0	19.0-22.0	0.045	0.03	
314	0.25	2.00	1.5-3.0	23.0-26.0	19.0-22.0	0.045	0.03	
316	0.08	2.00	1.00	16.0-18.0	10.0-14.0	0.045	0.03	2.0-3.0 Mo
316L	0.03	2.00	1.00	16.0-18.0	10.0-14.0	0.045	0.03	2.0-3.0 Mo
317	0.08	2.00	1.00	18.0-20.0	11.0-15.0	0.045	0.03	3.0-4.0 Mo
317L	0.03	2.00	1.00	18.0-20.0	11.0-15.0	0.045	0.03	3.0-4.0 Mo
321	0.08	2.00	1.00	17.0-19.0	9.0-12.0	0.045	0.03	5 × %C Ti min
329	0.10	2.00	1.00	25.0-30.0	3.0-6.0	0.045	0.03	1.0-2.0 Mo
330	0.08	2.00	0.75-1.5	17.0-20.0	34.0-37.0	0.04	0.03	
347	0.08	2.00	1.00	17.0-19.0	9.0-13.0	0.045	0.03	C
348	0.08	2.00	1.00	17.0-19.0	9.0-13.0	0.045	0.03	0.2 Cu[b, c]
384	0.08	2.00	1.00	15.0-17.0	17.0-19.0	0.045	0.03	

a. Single values are maximum unless indicated otherwise. b. (Cb + Ta) min – 10 × %C. c. Ta – 0.10% max.

Fig. 9-1. AISI (American Iron and Steel Institute) identifies stainless steel by a numbering system and specifies the composition of each type. Note the following symbols for the listed elements: Carbon (C), Manganese (Mn), Silicon (Si), Chromium (Cr), Nickel (Ni), Phosphorus (P), Sulfur (S).

corrosion. This steel cannot be hardened by heat treatment and is nonmagnetic. Such steels are commonly called AUSTENITIC stainless steels due to their grain structure.

Filler Materials

Various filler materials are used to join the alloys. For best results, the filler wire should match the base metal or base metal combinations. The filler wires used are shown in Fig. 9-2. When selecting stainless steel filler wire, use low carbon (LC) or extra low carbon (ELC) filler wire whenever specified. These wires are also used for welding when not specified.

Joint Preparation and Cleaning

The same procedures used on carbon steels may be used for stainless steels with one exception. DO NOT USE CARBON STEEL WIRE BRUSHES ON THE PARENT MATERIAL OR THE WELD. The carbon will be picked up by the base metal and absorbed into the weld, adversely affecting completed welds.

Preheat, Interpass, and Postheat Temperatures

The chromium nickel stainless steels do not require preheat, interpass, or postheat. These materials do not harden by cooling; they are always in the austenite condition. However, they are prone to cracking if overheated. Excess heat also causes CARBIDE PRECIPITATION in unstabilized materials. (Carbide precipitation is a condition where chromium and carbon within the grains migrate to the grain boundries. The grains are then left unprotected against corrosion. When this weld is used in some acids, the metal will be subject to intergranular corrosion and perhaps failure of the weldment.)

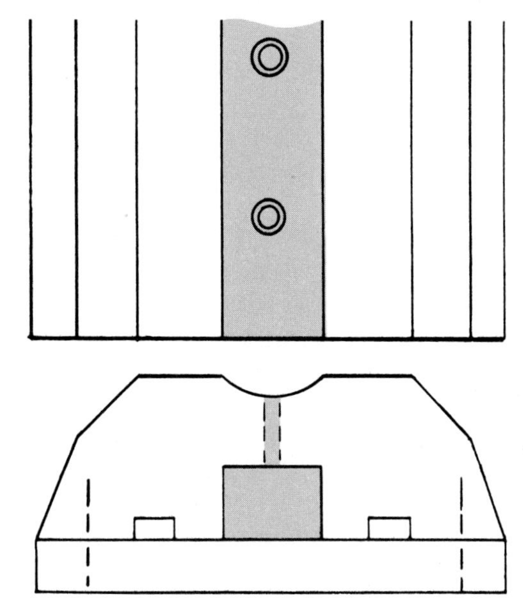

Fig. 9-3. Backup bar design for admitting gas to the penetration side of the weld joint.

Tooling

Nonmagnetic tooling must always be used to prevent magnetic arc blow. Stainless steel does not transfer heat as readily as carbon steel. Therefore, tooling must remove heat rapidly to prevent warpage and distortion. Copper is often used for tooling; it is a good, nonmagnetic heat conductor.

The root side of full penetration welds should have an inert backup gas. A copper strip machined with a groove to accept the drop through, as shown in Fig. 9-3, prevents oxidation of the metal from the atmosphere. It also assists in the formation of the penetration bead.

Weld Rod	308	308L	309	309Cb	310	310Cb	310Mo	312	316	316L	317	318	347	410	420	430
Base Material	201	201L	309	309	310	310Cb	310Mo	312	316	316L	317	316	347	410	420	430
	202	304L	442	442	442	442		501	301	321	301	318	301	308	309	309
	301	301	201	201	201			502	302	347	302	301	302	347	308	308
	302	302	202	202	202				304		304	302	304			308L
	304	304	301	301	301				317		305	304	304L			309
	305	305	302	302	302							305	318			309Cb
	405	321	304	304	304							316	321			310
	409	347	305	305	305							317	403			310Cb
	410											347	405			347
	430												409			
													410			
													430			
													501			
													502			

Fig. 9-2. Filler materials listed at the top of each column may be used to weld the materials listed in that column (either alone or in combination).

Another method of protecting the root side of the weld joint is to use a flux designed for this particular use. A flux is shown in Fig. 9-4. The flux powder is mixed with acetone or alcohol and made into a paste. It is applied to the root side of the joint with a brush. The alcohol or acetone will evaporate leaving the flux behind. A weld joint with the paste applied is shown in Fig. 9-5. During the welding operation, the flux melts and flows over the molten metal preventing oxidation. After welding, the remaining flux is removed by brushing the part in hot water or using a commercial acid cleaner.

Welding Procedures

Stainless steel procedures are listed in the reference section to set up the machines for both short and spray arc techniques. The forehand technique is recommended because it forms a flatter weld crown. The use of stringer beads when welding stainless steels reduces overall heat input and the possibility of cracking.

Fig. 9-4. Typical commercial flux is often used where backup bars or tooling cannot be used. (Golden Empire Corp.)

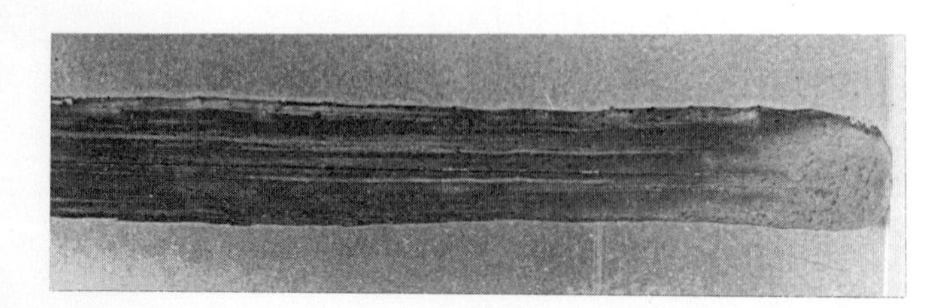

Fig. 9-5. The flux powder remains on the part after the fluid evaporates.

When using the short arc technique of welding, the gas mixture must be a TRI-MIX gas consisting of 90 percent helium, with seven and a half percent argon, and two and a half percent carbon dioxide.

A stainless steel tank ready for automatic welding with this gas mixture is shown in Fig. 9-6. Welding is quite rapid; therefore, the position of the torches is offset to control the crown height of the finished weld. A completed weld is shown in Fig. 9-7. Note that the sides of this weld are very dark. This is due to the reaction of the arc with the helium shielding gas.

When using the spray arc technique, the gas mixture must be either 98 percent argon with two percent oxygen or 99 percent argon with one percent oxygen.

Fig. 9-6. This stainless steel tank assembly uses a joggle weld joint design. Therefore, a backup gas or flux is not required to prevent oxidation of the weld root. (FluidTech Corp.)

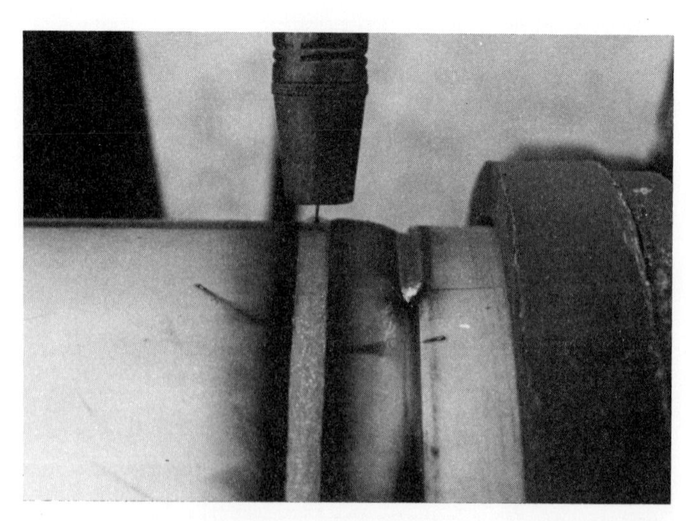

Fig. 9-7. The completed weld is very smooth with good bead contour. The soot may be removed by wire brushing or with a chemical cleaner. (FluidTech Corp.)

When making multiple pass welds, be sure to clean every bead with a stainless steel brush before making another pass.

Do not leave craters at the end of the weld. Cracks will form from the crater. To prevent this condition in manual welding, reverse the direction of the torch and stop the weld on the full cross section.

Stainless steel is cleaned after welding with a commercial cleaner like in Fig. 9-8.

Fig. 9-8. Commercial stainless steel weld cleaner. (Arcal Chemical, Inc.)

SPOT WELDING STAINLESS STEEL

Chromium nickel stainless steels are welded with the same equipment as carbon steels. Whenever possible, place the thinnest material on top. Since these materials have an oxide film on the surface, always clean the mating areas where the weld is to be located to bright metal before assembling the components. See Fig. 8-2. Follow these steps when welding stainless steel:

1. Match the filler wire with the base material.
2. Match the gas mixture with the mode of welding.
3. Apply flux or backup gas where full penetration welds are used.
4. Apply tooling where required to prevent distortion.

PLUG AND SLOT WELDING STAINLESS STEEL

Use the same procedures for seam welding and spot welding. The weld techniques used for these welds are the same as those in carbon steel welds.

Weld Defects

Do not feel frustrated when some of your welds contain defects. As your skills progress with practice, you will learn to avoid these mistakes. Chapter 11 discusses weld defects that are common when making various types of welds, regardless of the material. Refer to this chapter to learn the corrective action for the various defects you may encounter.

REVIEW QUESTIONS—Chapter 9

1. Chromium nickel stainless steels are commonly called ________ stainless steels. This word also describes the grain structure of these materials.
2. The chromium nickel stainless steels are identified by the ________ ________ ________ ________.
3. This group of metals has very good ________ resistance.
4. The letters LC or ELC identify that the welding material must have a ________ ________ or ________ ________ ________ content.
5. When brushing a stainless steel base material or weld, always use a ________ ________ wire brush.
6. When welding stainless steel materials, ________ is not needed. If the material is overheated during the welding operation, ________ may result.
7. Nonmagnetic tooling reduces the possibility of ________ ________ ________ during welding.
8. When welding stainless steel longseam or fillet welds, the ________ welding technique is used to form a flatter weld crown.
9. The root side of full penetration welds will be ________ if a gas or flux is not used during welding.
10. Short-arc welding on stainless steels uses a gas combination called ________-________.
11. Short-arc welds made with this gas have a sooty deposit next to the completed weld. Part of this mixture, ________, is the main reason for this condition.
12. Spray-arc welds on stainless steels use a mixture of argon and ________.
13. Craters occur at the ends of stainless steel welds, and if not filled properly ________ can occur.
14. A special ________ is used to remove heat lines along the weld.
15. Welding procedures for stainless steels are basically the same as for ________ steels.

Chapter 10

WELDING ALUMINUM

After studying this chapter, you will be able to:

- List eight characteristics of aluminum.
- Name the two basic groups of aluminum alloys and list many of them by number.
- Match the proper filler wire with various aluminum alloys for welding.
- Choose the correct filler wire to produce desired properties in a weld.
- Apply tooling to a joint for welding.
- Prepare an aluminum joint for welding.
- Demonstrate proper techniques in welding aluminum.

Aluminum is a nonferrous (no iron) metal readily welded with the GMAW process. Some major characteristics of aluminum include:

1. Thermal conductivity.
2. Electrical conductivity.
3. Ductility at subzero temperatures.
4. Light weight.
5. High resistance to corrosion.
6. Nonsparking.
7. Nontoxic.
8. Does not change color when heated.

Pure aluminum melts at approximately 1200°F. Aluminum alloys will melt at approximately 900° to 1200°F.

An oxide film which forms on the surface of aluminum provides good corrosion resistance. However, this film must be removed prior to welding to produce quality welds.

Base Materials

Aluminum materials are made in two groups, hardenable and nonhardenable. The hardenable group may be welded in the soft or hardened condition. After welding, strength levels may be changed as desired by further heat treatment. Hardenable and nonhardenable wrought alloys are shown in Fig. 10-1.

Non-Heat-Treatable (alloys normally cold-worked)	Heat-Treatable (alloys normally heat-treated)
1060	2011
1100	2014
3003	2017
3004	2018
4043	2024
5005	2025
5050	2117
5052	2218
5056	2618
5083	4032
5086	6053
5184	6061
5252	6063
5257	6066
5357	6101
5454	6151
5456	7039
5557	7075
5657	7079
	7178

Fig. 10-1. Hardenable and nonhardenable aluminum base materials.

Filler Materials

The filler material choice for welding various aluminum alloys is important. The metal produced in the weld is a mixture of the parent metal and the filler material. In addition to resisting cracking and corrosion, a typical weld must be strong and DUCTILE. (Ductility is a property of a material to deform permanently, or to exhibit plasticity without breaking while under load.) The correct choice of a filler wire or alloy will eliminate or reduce low ductility in aluminum welds.

Maximum weld quality can only be obtained if the filler material is clean and of high quality. The filler material should be protected from oil, dirt, grease, or other types of contamination. When not in use in the shop, filler wire should be removed from the machine and stored in a clean, dry area. The use of contaminated wire with GMAW is one of the major causes of poor weld quality.

The proper filler wire for welding various alloys is shown in Fig. 10-2.

The proper filler wire for achieving specific properties of completed welds is found in Fig. 10-3.

Base Metal	Filler Alloys[1]		Base Metal	Filler Alloys[1]	
	Preferred for Maximum As-welded Tensile Strength	Alternate Filler Alloys For Maximum Elongation		Preferred for Maximum As-welded Tensile Strength	Alternate Filler Alloys For Maximum Elongation
EC	1100	EC/1260	5086	5183	5183
1100	1100/4043	1100/4043	5154	5356	5183/5356
2014	4145	4043/2319[3]	5357	5554	5356
2024	4145	4043/2319[3]	5454	5554	5356
2219	2319	(4)	5456	5556	5183
3003	5183	1100/4043	6061	4043/5183	5356[2]
3004	5554	5183/4043	6063	4043/5183	5183[2]
5005	5183/4043	5183/4043	7039	5039	5183
5050	5356	5183/4043	7075	5183	—
5052	5356/5183	5183/4043	7079	5183	—
5083	5183	5183	7178	5183	(4)

The above table shows recommended choices of filler alloys for welds requiring maximum mechanical properties. For all special services of welded aluminum, inquiry should be made of your supplier.

1. Data shown are for "O" temper.
2. When making welded joints in 6061 or 6063 electrical conductor in which maximum conductivity is desired, use 4043 filler metal. However, if strength and conductivity both are required, 5356 filler may be used and the weld reinforcement increased in size to compensate for the lower conductivity of the 5356 filler metal.
3. Low ductility of weldment is not appreciably affected by filler used. Plate weldments in these base metal alloys generally have lower elongations than those of other alloys listed in this table.

Fig. 10-3. Filler materials used to obtain specific properties of completed welds.

Base Metal	6070	6061, 6063, 6101, 6151, 6201, 6951	5456	5454	5154, 5254[a]	5086	5083	5052, 5652[a]	5005, 5050	3004, Alc. 3004	2219	2014, 2024	1100, 3003, Alc. 3003	1060, EC
1060, EC	ER4043[h]	ER4043[h]	ER5356[c]	ER4043[e,h]	ER4043[e,h]	ER5356[c]	ER5356[c]	ER4043[i]	ER1100[c]	ER4043	ER4145	ER4145	ER1100[c]	ER1100
1100, 3003, Alclad 3003	ER4043[h]	ER4043[h]	ER5356[c]	ER4043[e,h]	ER4043[e,h]	ER5356[c]	ER5356[c]	ER4043[e,h]	ER4043[e]	ER4043[e]	ER4145	ER4145	ER1100[c]	
2014, 2024	ER4145	ER4145									ER4145[g]	ER4145[g]		
2219	ER4043[f,h]	ER4043[f,h]	ER4043	ER4043[h]	ER4043[h]	ER4043[i]	ER4043	ER4043[i]	ER4043	ER4043	ER2319[c,f,h]			
3004, Alclad 3004	ER4043[e]	ER4043[b]	ER5356[e]	ER5654[b]	ER5654[b]	ER5356[e]	ER5356[e]	ER4043[e,h]	ER4043[e]	ER4043[e]				
5005, 5050	ER4043[e]	ER4043[b]	ER5356[e]	ER5654[b]	ER5654[b]	ER5356[e]	ER5356[e]	ER4043[e,h]	ER4043[d]					
5062, 5652[a]	ER5356[b,c]	ER5356[b,c]	ER5356[b]	ER5654[b]	ER5654[b]	ER5356[e]	ER5356[e]	ER5654[a,b,c]						
5083	ER5356[e]	ER5356[e]	ER5183[e]	ER5356[e]	ER5356[e]	ER5356[e]	ER5183[e]							
5086	ER5386[e]	ER5356[e]	ER5356[e]	ER5356[b]	ER5356[b]	ER5356[e]								
5154, 5254[a]	ER5356[b,c]	ER5356[b,c]	ER5356[b]	ER5654[b]	ER5654[a,b]									
5454	ER5356[b,c]	ER5356[b,c]	ER5356[b]	ER5554[a,b]										
5456	ER5356[e]	ER5356[e]	ER5556[e]											
6061, 6063, 6101, 6201, 6151, 6951	ER4043[b,h]	ER4043[b,h]												
6070	ER4043[e,h]													

Where no filler metal is listed, base metal combination is not recommended for welding.

a. Base metal alloys 5652 and 5254 are used for hydrogen peroxide service. ER5654 filler metal is used for welding both alloys for low-temperature service (150 F and below).
b. ER5183, ER5356, ER5554, ER5556, and ER5654 may be used.
c. ER4043 may be used.
d. Filler metal with same analysis as base metal is sometimes used.
e. ER5183, ER5356, or ER5556 may be used.
f. ER4145 may be used.
g. ER2319 may be used.
h. ER4047 may be used.
i. ER1100 may be used.

Fig. 10-2. Wrought aluminum alloy filler material selection.

Joint Preparation and Cleaning

Joint edges prepared by the plasma arc cutting process or the carbon arc gouging process form a heavy oxide film on the surface. This surface should be thoroughly cleaned prior to welding. This prevents DROSS and porosity in the final weld. (Dross is oxidized metal or impurities within the parent metal or weld.) This heavy oxidization should be removed by machining or sanding. DO NOT USE GRINDING WHEELS TO CLEAN THE OXIDE FILM FROM ALUMINUM.

Joint edges prepared by the shearing process should be sharp without tearing or ridges. Where these conditions exist, dirt or oil may become entrapped and result in faulty welds.

Preweld cleaning consists of removing oxidation from weld joint edges and the immediate weld area. This can be done by:

1. Chemical cleaning.
   A. Commercial degreasers.
   B. Commercial compounds, Fig. 10-4.
2. Mechanical cleaning.
   A. Filing or scraping.
   B. Abrasive pads or sanding with abrasive wheels.
   C. Stainless steel brushes. (DO NOT USE CARBON STEEL WIRE BRUSHES.)

After cleaning is completed, but before welding, wash the weld joint with alcohol or acetone. Allow to dry before welding. DO NOT WELD ON A WET SURFACE.

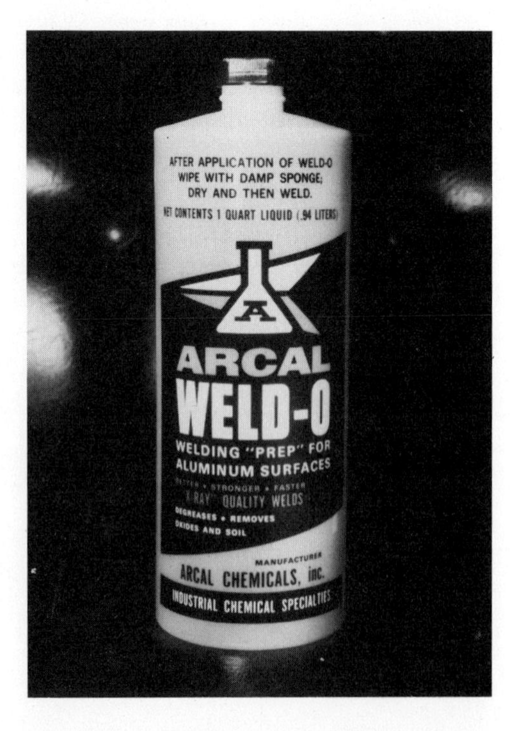

Fig. 10-4. Commercial aluminum cleaner. (Arcal Chemicals, Inc.)

Tooling

Back up tooling is needed when welding full penetration groove welds without a backup bar. Care should be taken that molten aluminum does not drop through the weld joint into the atmosphere. This prevents contamination and insures that the puddle will flow properly. Tooling with a groove machined into the face, with or without argon gas, protects the weld metal. An example of this type of tooling is located in Fig. 10-5.

Welding Procedures

The procedures for spray arc-welding various types and thicknesses of aluminum are given in the reference section. Short-arc welding may be used for welding of thin aluminum. However, short-arc welds are prone to contain porosity due to the fast freezing of the molten pool. For this reason, the spray-arc welding mode should be used whenever possible.

Argon gas is the main gas used for welding with both modes. However, when welding heavier sections, helium should be added. This yields a hotter puddle, better penetration, and less porosity.

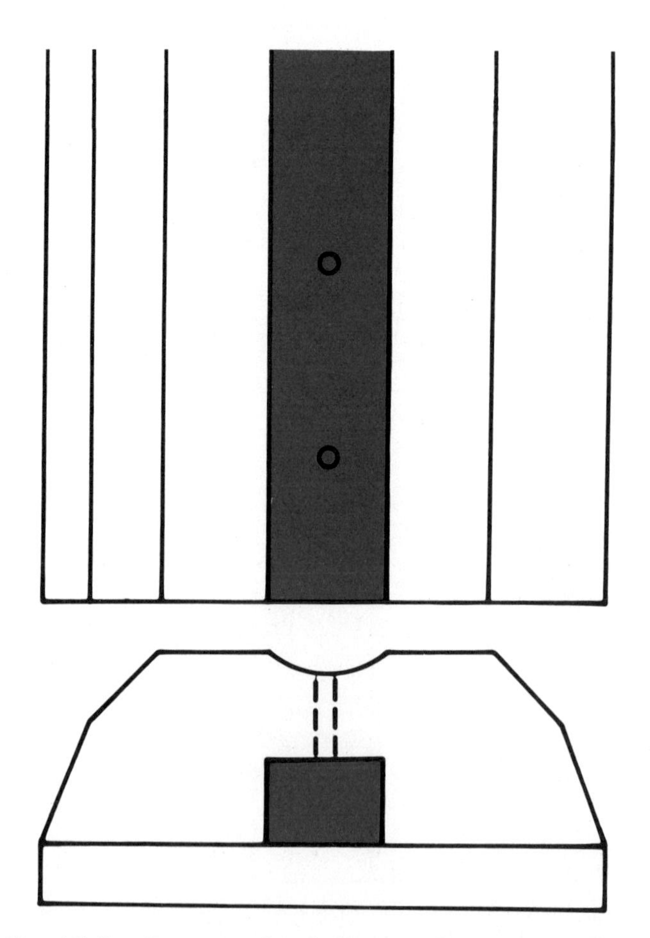

Fig. 10-5. Argon gas is admitted to the lower section of the tool and the gas flows through the holes to the grooved area. The gas then shields the penetration during welding and prevents the formation of oxides. This also assists in forming the root bead and prevents oxide folds.

When using the spray arc mode on fillet or seam welds, the forehand technique will have good cleaning action in the weld joint. Fig. 10-6 shows a weld made with both backhand and forehand techniques and the final results.

When making multipass welds, always clean the previous weld with a stainless steel wire brush to remove the oxide film.

Do not stop the weld at the end of a joint and leave a crater like the one in Fig. 10-7. Aluminum is prone to cracking when the cross section of the weld is thin. Using the same weld travel torch angle, always move the torch back on the full cross section of the weld before stopping. Where this cannot be done, use a tab of aluminum metal to prevent cracking. A tab is used in Fig. 10-8.

SPOT WELDING, PLUG AND SLOT WELDING

Use the procedures outlined in this chapter and specified in the reference section for these types of welds. Welding techniques for these joints are the same as those used with carbon and stainless steel.

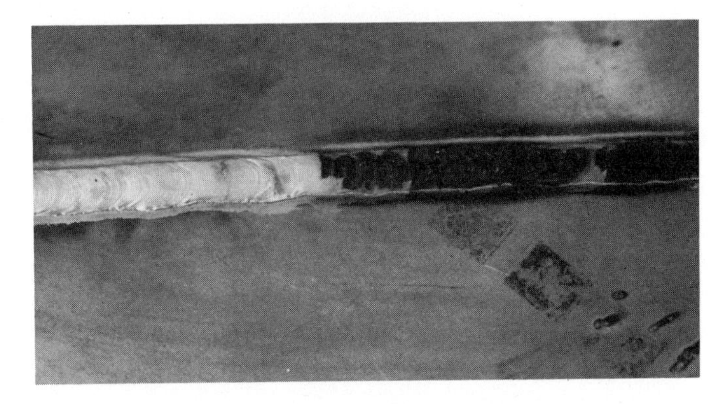

Fig. 10-6. The dark colored section of the weld was made with the backhand welding technique. (Note the smoke and soot on the weld and weld area.) The clean section of the weld was made with the forehand welding technique. The weld is clean. The gray area adjacent to the weld indicates the cleaning action of the reverse polarity welding current.

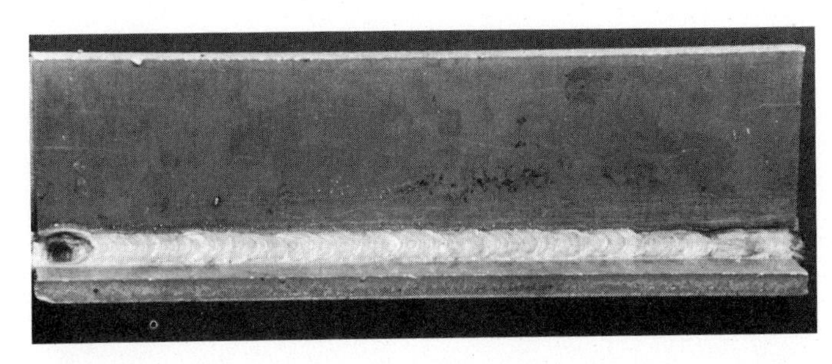

Fig. 10-7. The crater at the end of the weld was formed when the welding stopped. This part of the weld is very porous and has cracks throughout the crater.

BURNBACK

Burnback of the welding wire can be a very serious and costly problem when welding aluminum. Although burnback may occur in GMAW during any mode with any type of wire, this problem is greater with welding aluminum. This is due to its very low melting point (1200°F). Fig. 10-9 illustrates how the molten metal flows over the tip end and possibly inside the wire hole. Burnback could easily ruin a tip. The following list suggests how to minimize this condition:

1. When setting up for a weld operation, use a close arc voltage and then adjust to the desired voltage.
2. When setting up for a weld operation, use a higher wire feed speed and then adjust to the desired value.
3. Use very short arc times and watch the arc very closely until the correct values are established for the desired weld.

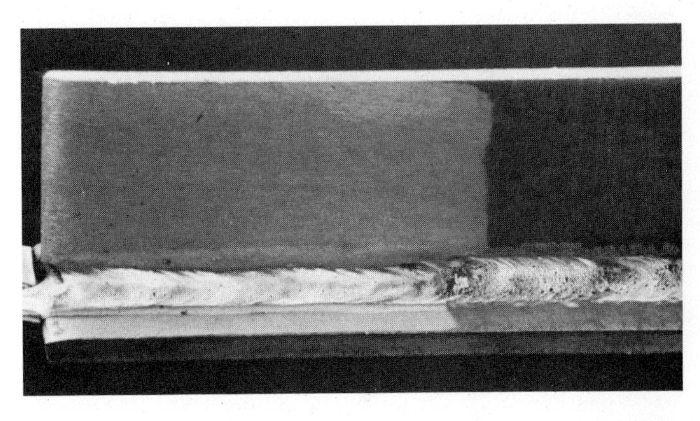

Fig. 10-8. This fillet weld has a run off tab in place to prevent craters at the end of the weld. Part of the weld area was cleaned before welding. Note the weld appearance between the two areas.

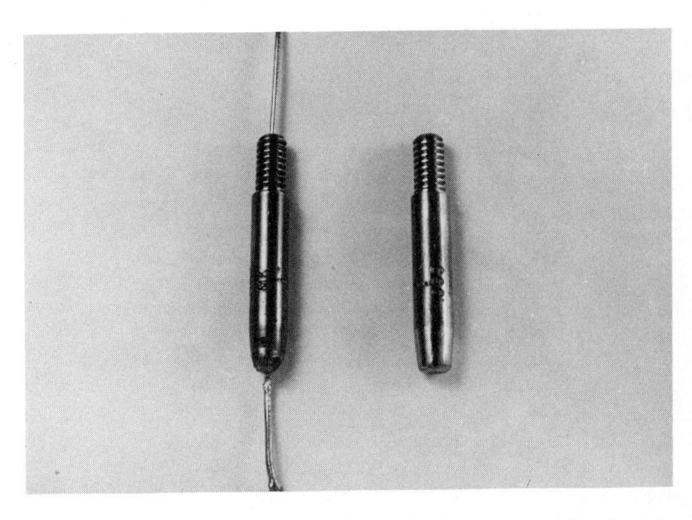

Fig. 10-9. Left. Aluminum filler wire has burned back over the end of the tip. This tip cannot be repaired to the original condition of the tip shown on the right.

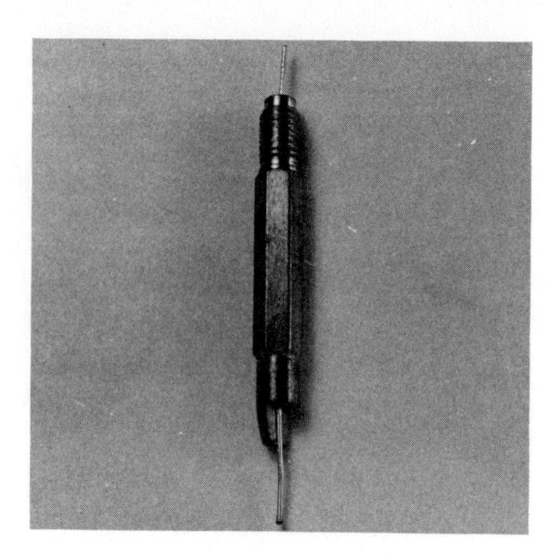

Fig. 10-10. The tip is notched to give access to the wire end. The wire may then be pried away from the tip for removal. The tip end may be lightly filed to remove any extra metal.

4. Watch your wire supply very closely. If the supply runs out during a weld, burnback will occur and possibly ruin the contact tip. Always know how much wire is on the spool.

The contact tip shown in Fig. 10-10 has been specially machined to assist in the clearing of the burnback.

Weld Defects

Do not feel frustrated when some of your welds contain defects. As your skills progress with practice, you will learn to avoid these mistakes. Chapter 11 discusses weld defects that are common when making various types of welds, regardless of the material. Refer to this chapter to learn the corrective action for the various defects you may encounter.

REVIEW QUESTIONS—Chapter 10

1. List eight characteristics of aluminum.
2. What is a nonferrous material? Is aluminum nonferrous?
3. The oxide film on the surface of aluminum forms during the rolling and cooling period in manufacturing. This film aids in preventing ________ of the base material during use.
4. Why must the oxide film be removed before welding aluminum?
5. The choice of filler wire in the weld joint will have a definite effect on the completed weld's ________.
6. The use of ________ filler material will have adverse effects on the quality of the completed weld.
7. Heavy oxide scale on the surface of a thermal cut joint should be removed by ________ or sanding.
8. ________ ________ should never be used to prepare aluminum weld joints.
9. Commercial compounds or acids and ________ ________ wire brushes should be used to remove oxide films from joints or welds.
10. Final cleaning before welding may be done with ________ or ________ which remove contaminants.
11. What mode and technique are used most frequently in welding aluminum with this process?
12. This technique has a ________ ________ which removes surface oxides.
13. How does adding helium to the shielding gas affect weld quality?
14. How are craters avoided at the end of a weld pass?
15. Where possible, ________ are used to reduce cracking at the end of the weld.

Chapter 11

WELD DEFECTS AND CORRECTIVE ACTION

After studying this chapter, you will be able to:

■ Identify common groove, fillet, plug, and spot weld defects and take action to correct them.

■ Make proper use of weld inspection tools.

During the GMAW process, defects occur which may affect the quality of the weld. As a result, the weld could fail under stress. Most defects are caused by improper welding procedure. Still others fail from lack of welder skill. Even though the process is semiautomatic, and the amperage and voltage are controlled by the machine, these defects can occur for many reasons.

Each listed defect will refer you to a figure which, in addition to having a picture, describes the defect. By using the corrective action listed with each, these defects can be avoided.

GROOVE WELD DEFECTS

1. Lack of penetration. Fig. 11-1.
2. Lack of fusion. Fig. 11-2.
3. Overlap. Fig. 11-3.
4. Undercut. Fig. 11-4.
5. Convex bead. Fig. 11-5.
6. Craters. Fig. 11-6.
7. Cracks. Fig. 11-7.
8. Porosity. Fig. 11-8.
9. Linear porosity. Fig. 11-9.
10. Burnthrough. Fig. 11-10.
11. Whiskers. Fig. 11-11.
12. Excessive penetration and icicles. Fig. 11-12.
13. Spatter. Fig. 11-13.

FILLET WELD DEFECTS

1. Lack of penetration. Fig. 11-14.
2. Lack of fusion. Fig. 11-15.
3. Overlap. Fig. 11-16.
4. Undercut. Fig. 11-17.
5. Convexity. Fig. 11-18.
6. Craters. Fig. 11-19.
7. Cracks. Fig. 11-20.
8. Burnthrough. Fig. 11-21.
9. Porosity. Fig. 11-22.
10. Linear porosity. Fig. 11-23.

PLUG WELD DEFECTS

1. Lack of penetration. Fig. 11-24.
2. Excessive penetration. Fig. 11-25.
3. Cracks. Fig. 11-26.
4. Porosity. Fig. 11-27.
5. Overlap. Fig. 11-28.
6. Craters. Fig. 11-29.

SPOT WELD DEFECTS

1. Lack of penetration. Fig. 11-30.
2. Excessive penetration. Fig. 11-31.
3. Porosity. Fig. 11-32.
4. Cracks. Fig. 11-33.

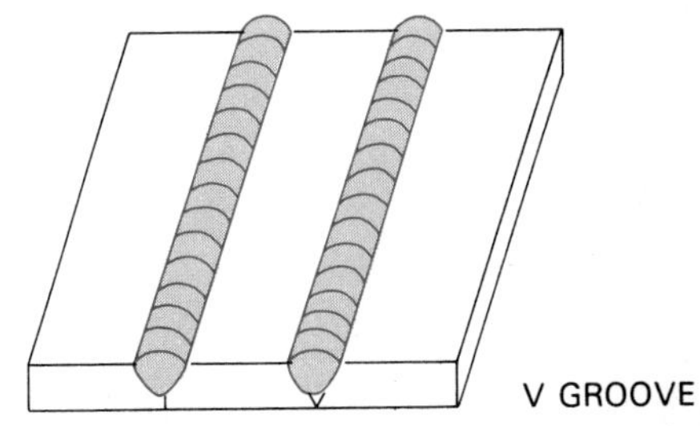

Fig. 11-1. LACK OF PENETRATION (Incomplete Penetration) describes a weld not sinking properly into the weld joint. TO CORRECT: open the groove angle, decrease the root face, increase root opening, increase amperage, decrease voltage, decrease travel speed, change torch angle, decrease stickout, or keep the arc on the leading edge of the molten pool.

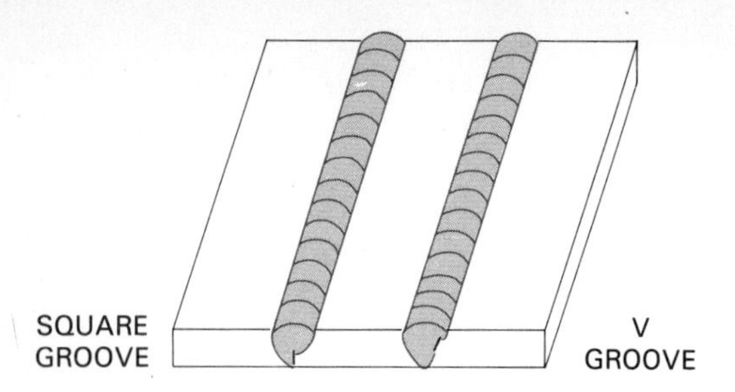

Fig. 11-2. LACK OF FUSION (Cold Shut) often occurs in multiple pass welds where the passes do not fuse. TO CORRECT: clean weld joint before welding, remove any oxides from previous welds, open groove angle, decrease root face, increase root opening, increase amperage, decrease voltage, decrease travel speed, change torch angle, decrease stickout, or keep the arc on the leading edge of the molten pool.

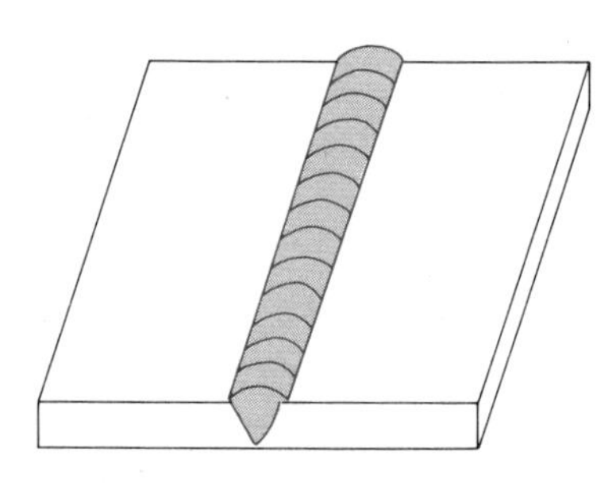

Fig. 11-3. OVERLAP is weld metal that has flowed over the edge of the joint and improperly fused with the parent metal. TO CORRECT: clean edge of weld joint, remove oxides from previous welds, reduce size of bead, or increase travel speed.

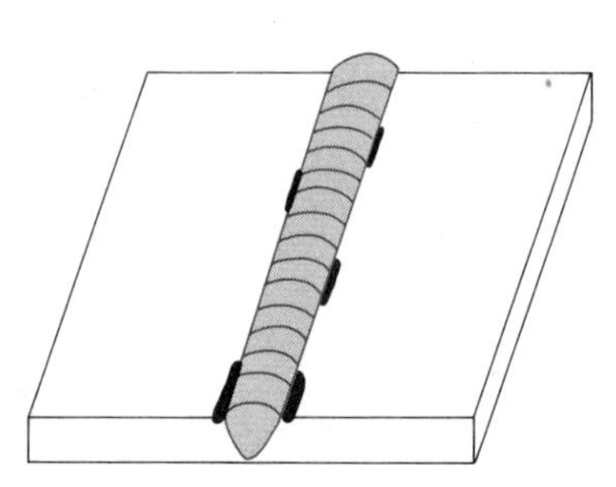

Fig. 11-4. UNDERCUT is lack of filler material at the toe of the weld metal. TO CORRECT: decrease travel speed, increase dwell time at the edge of the joint on wash beads, decrease voltage, decrease amperage, or change torch angle.

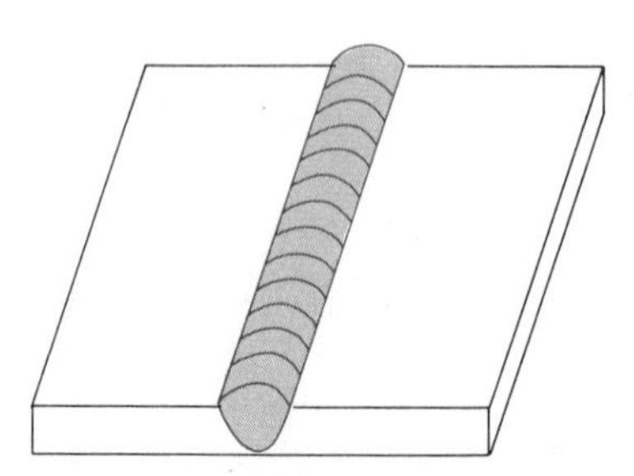

Fig. 11-5. CONVEX CROWN refers to a poorly formed weld crown that is peaked in the center. TO CORRECT: change torch angle, increase current, decrease stickout, or use a wash bead technique with a dwell time at the edge of the joint.

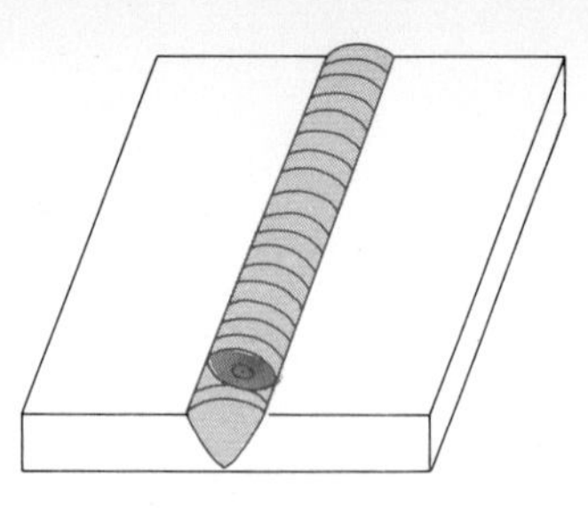

Fig. 11-6. CRATERS form when welding is stopped and the weld shrinks below the full cross section of the weld. TO CORRECT: do not stop welding at the end of the joint (use tabs), or using the same weld travel torch angles, move the torch back on the full section of the weld before stopping.

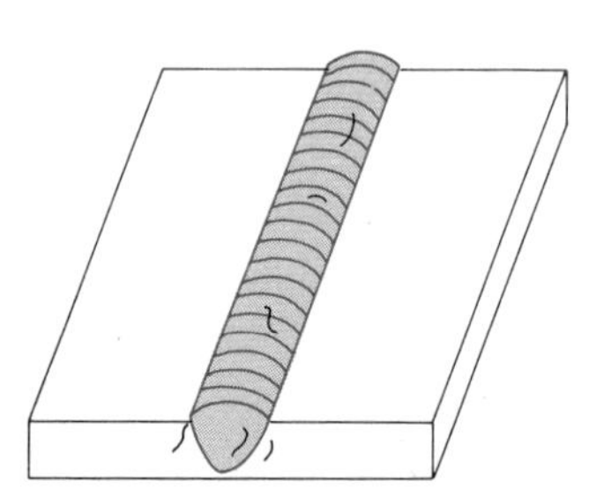

Fig. 11-7. CRACKS may occur anywhere in the weld or weld zone. Causes are weld and parent metal cooling stresses. TO CORRECT: Use a wire with a lower tensile strength or a different chemistry, increase joint preheat to slow weld cooling rate, allow joint to expand and contact during heating and cooling, or increase the size of the weld.

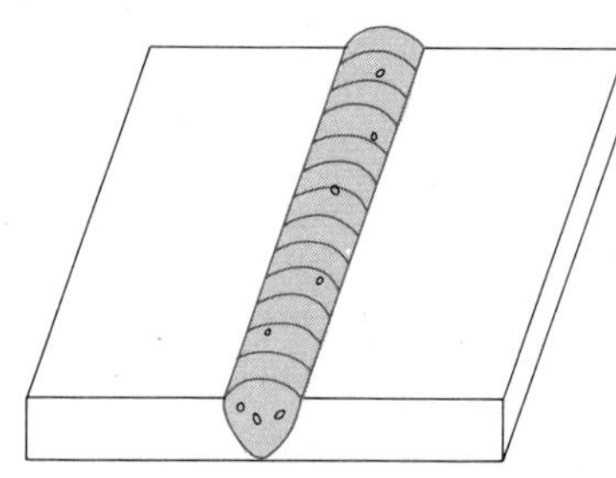

Fig. 11-8. POROSITY may form anywhere throughout the weld and is caused by entrapped gas which did not have enough time to rise through the melt to the surface. TO CORRECT: remove all heavy rust, paint, oil, or scale on the joint prior to welding, remove oxide film from previous weld beads or layers of weld, check shielding gas flow, protect weld area from wind gusts which may reduce shielding gas effect, remove spatter from interior of gas nozzle which may deflect gas flow, check gas hoses for leaks, and check gas supply for possible contamination.

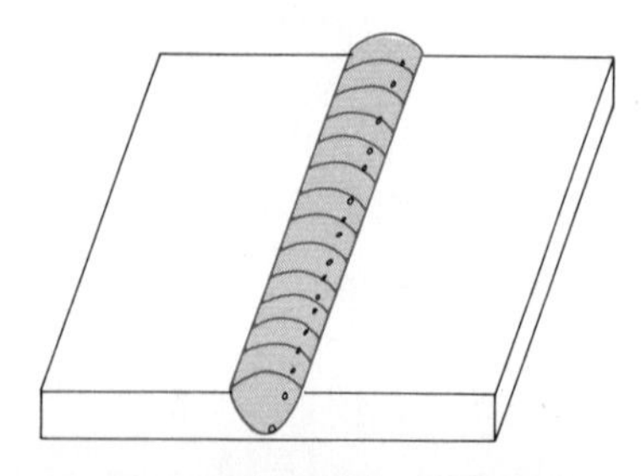

Fig. 11-9. LINEAR POROSITY usually forms in a line along the root of the weld at the center of the joint where penetration is very shallow. TO CORRECT: be sure root faces are clean, increase current, decrease voltage, decrease stickout, or slow down travel speed.

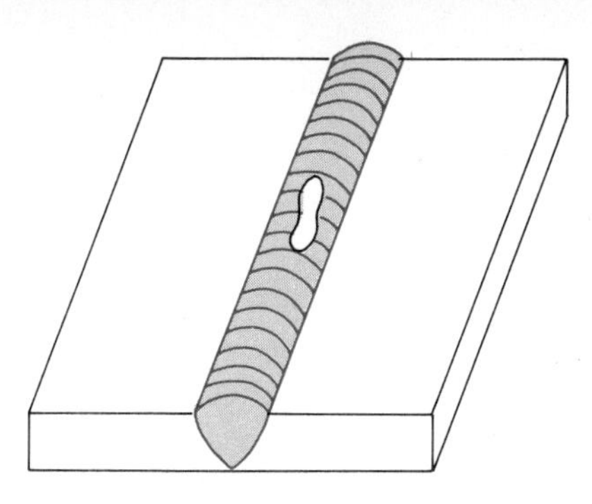

Fig. 11-10. BURNTHROUGH occurs where a gap may exist in the weld joint or where metal gets thinner. TO CORRECT: decrease current, increase voltage, increase stickout, increase travel speed, or decrease root opening.

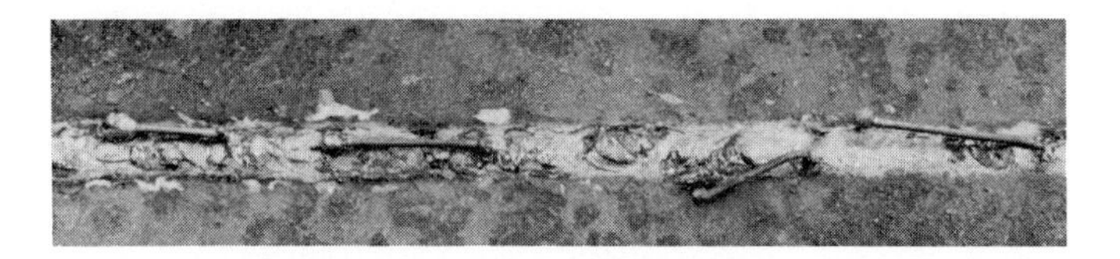

Fig. 11-11. WHISKERS are pieces of weld wire that extend through the weld joint. This usually occurs where there is a root spacing which the wire moves into. TO CORRECT: change weld technique to keep the arc on solidified metal and parent metal. Do not allow the electrode to move into the root opening.

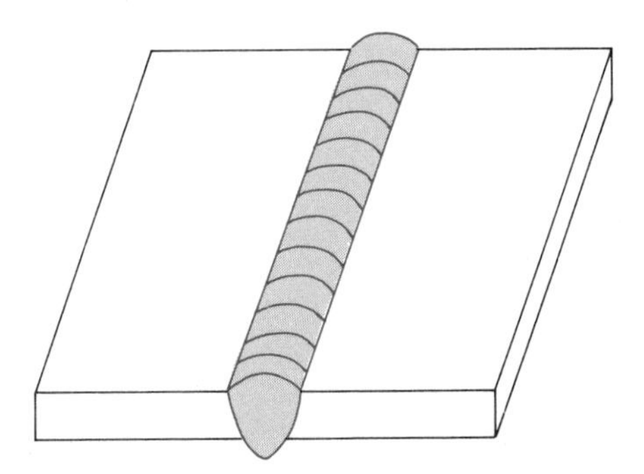

Fig. 11-12. EXCESSIVE PENETRATION and ICICLES are caused by extension of the weld below the root. TO CORRRECT: decrease root opening, increase root face, increase travel speed, decrease amperage, increase voltage, change torch angle, or increase stickout.

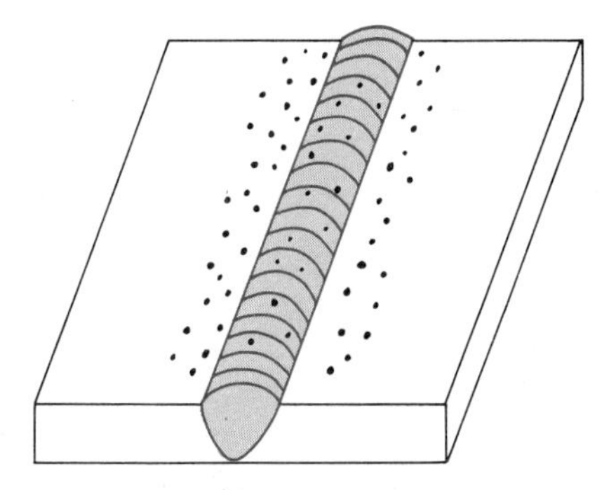

Fig. 11-13. EXCESSIVE SPATTER may form on the weld and parent metal. TO CORRECT: do not use carbon dioxide for welding on steel, change torch angle, or use a pull technique if possible.

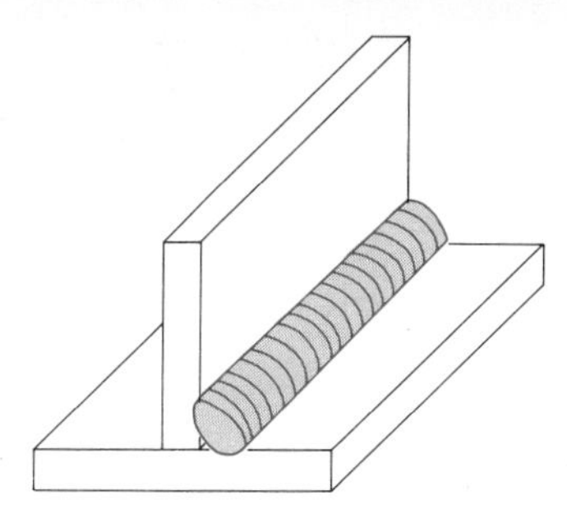

Fig. 11-14. LACK OF PENETRATION is insufficient weld metal intersection. TO CORRECT: change torch angle, increase amperage, decrease voltage, change to a gas which allows a hotter weld, decrease size of weld bead deposit, make stringer beads. Do not make weave beads on root passes.

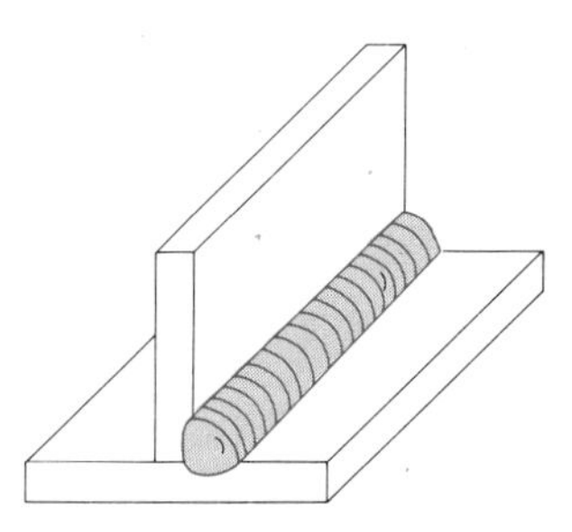

Fig. 11-15. LACK OF FUSION usually occurs in multiple pass welds where the passes or layers do not fuse. TO CORRECT: remove oxides and scale from previous weld passes, increase amperage, decrease voltage, decrease travel speed, change torch angle, decrease stickout, and keep the arc on the leading edge of the molten pool.

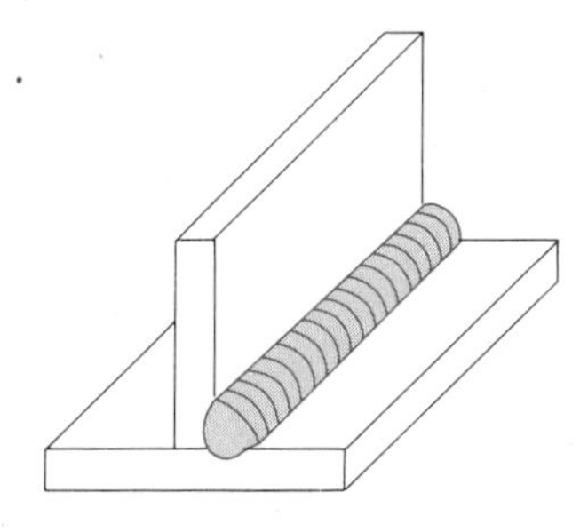

Fig. 11-16. OVERLAP usually occurs in horizontal fillet welds when too much weld is placed on the bottom pass in a multiple pass weld. TO CORRECT: reduce the size of the weld pass, reduce amperage, change torch angle, or increase travel speed.

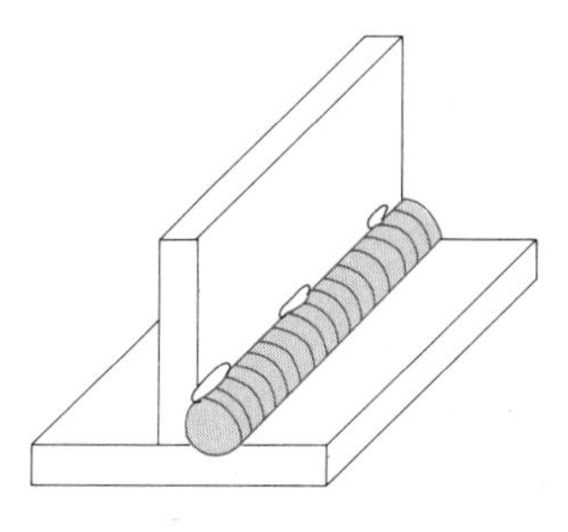

Fig. 11-17. UNDERCUT usually occurs at the top of the weld bead in horizontal fillet welds. TO CORRECT: make a smaller weld, a multiple pass weld, change torch angle, use a smaller diameter electrode, decrease amperage, or decrease voltage.

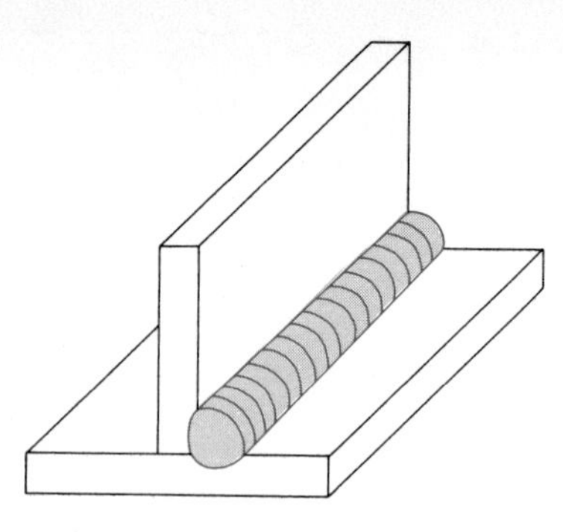

Fig. 11-18. CONVEXITY has a high crown. TO CORRECT: reduce arc voltage or amperage, decrease stickout, or change torch angle.

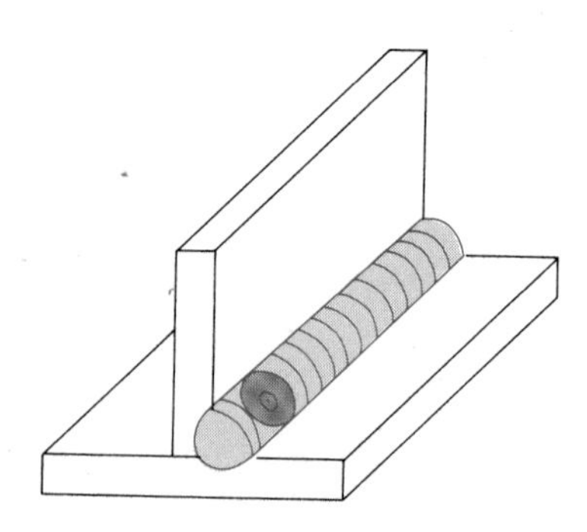

Fig. 11-19. CRATERS are formed when weld metal shrinks below the full cross section of the weld. TO CORRECT: do not stop welding at the end of the joint (use tabs), or using the same weld travel torch angle, move the torch back on the full cross section before stopping.

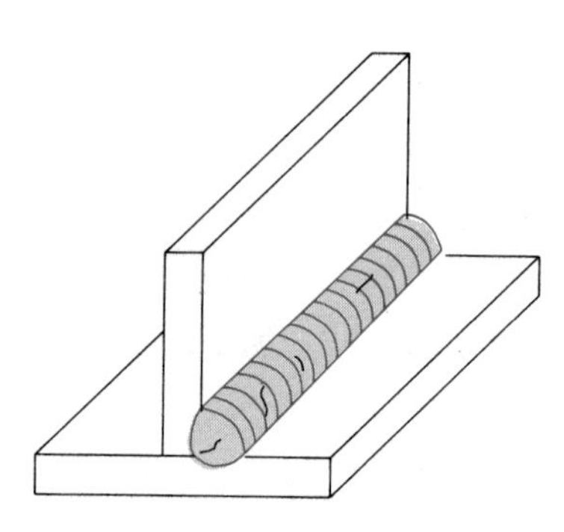

Fig. 11-20. CRACKS occur in fillet welds just as they do in groove welds. TO CORRECT: use suggestions listed in groove weld cracks.

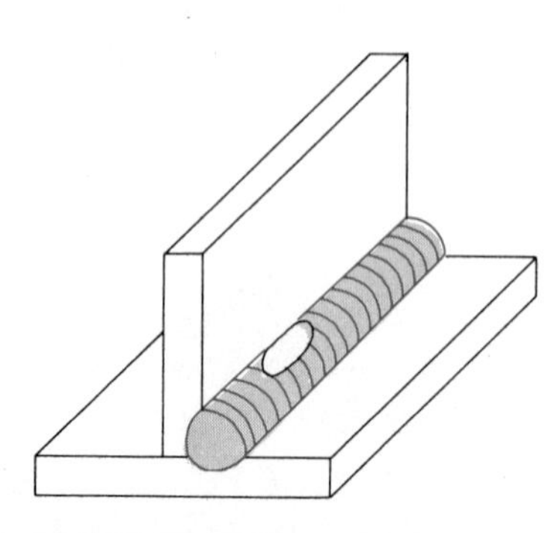

Fig. 11-21. BURNTHROUGH usually occurs in fillet welds on thin metals. TO CORRECT: decrease amperage, increase travel speed, or change torch angle.

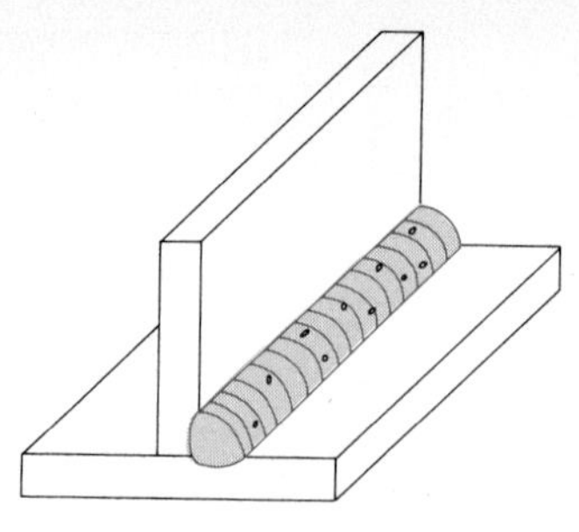

Fig. 11-22. POROSITY may form anywhere in the weld. It is entrapped gas which did not have enough time to rise through the melt to the surface. TO CORRECT: use suggestions listed in groove weld porosity.

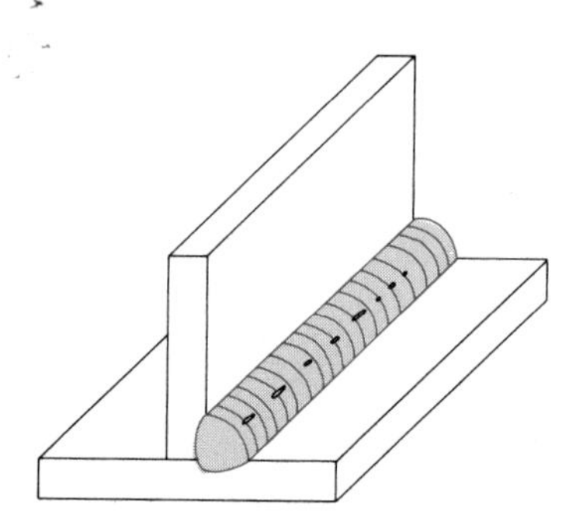

Fig. 11-23. LINEAR POROSITY usually forms along the root of the joint interface. TO CORRECT: use suggestions listed under groove weld linear porosity.

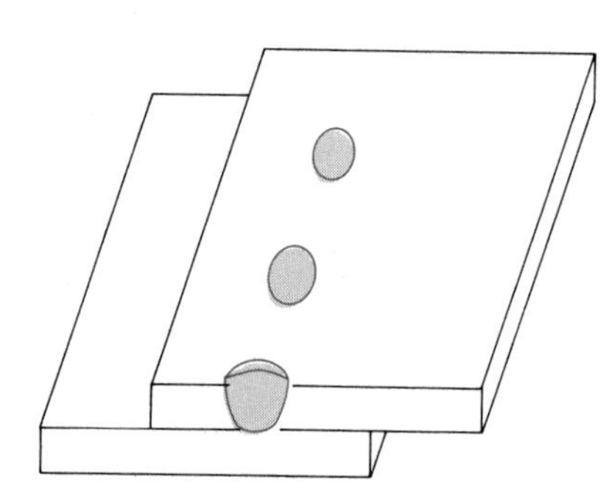

Fig. 11-24. LACK OF PENETRATION occurs when the weld metal does not reach the proper depth in the bottom plate. TO CORRECT: increase amperage, decrease voltage, decrease stickout, or start weld in the center of plug hole, and fill it in a circular pattern.

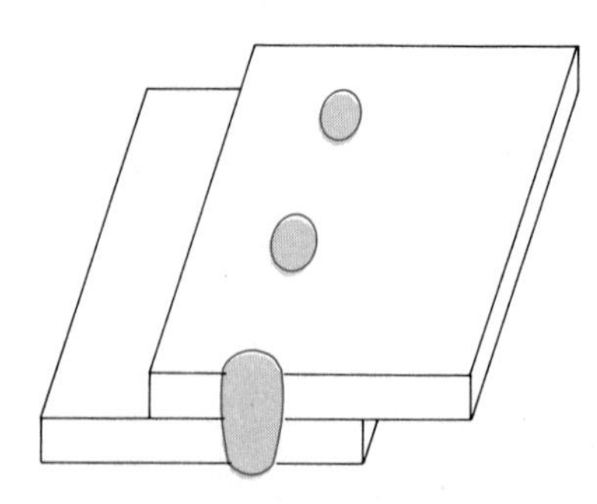

Fig. 11-25. EXCESSIVE PENETRATION occurs when the root of the weld extends too far below the bottom sheet of the assembly. TO CORRECT: decrease amperage, increase voltage, increase stickout, or move the torch in a circular pattern, and shorten weld operation.

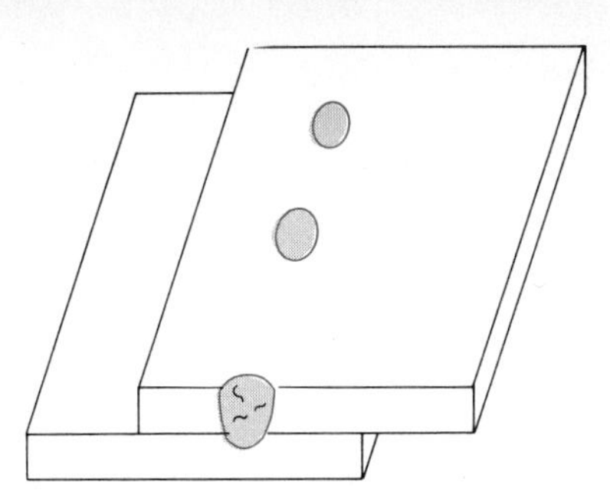

Fig. 11-26. CRACKS occur in plug welds at the center of the weld nugget due to rapid cooling of weld metal. TO CORRECT: use an electrode of a lower tensile strength or a different chemistry, increase preheat to slow down the cooling rate, or increase the size of the weld.

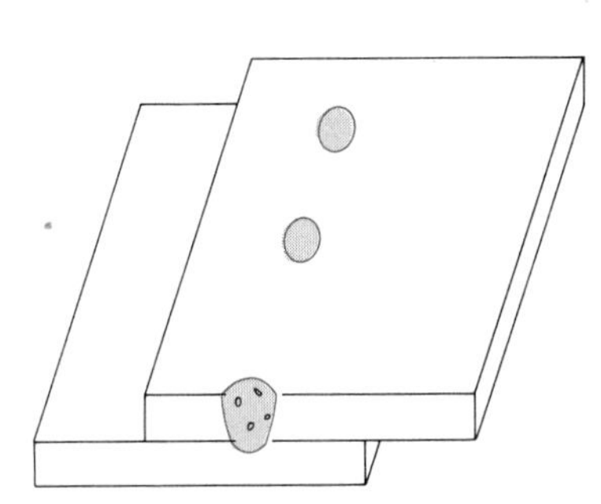

Fig. 11-27. POROSITY may form anywhere throughout the weld. It is caused by entrapped gas which did not have enough time to rise through the melt to the surface. TO CORRECT: use suggestions listed under groove weld porosity.

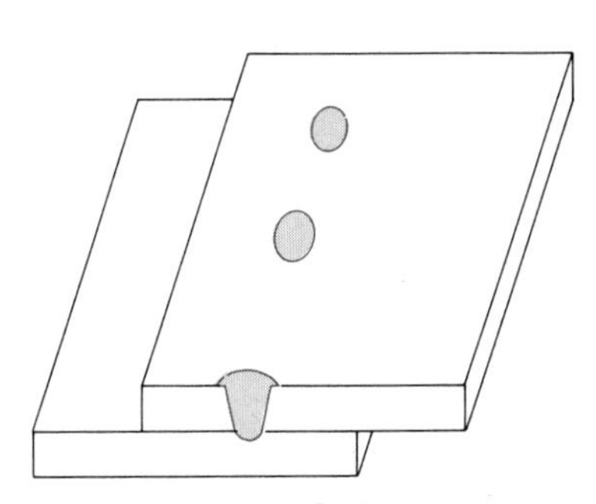

Fig. 11-28. OVERLAP occurs at the surface of the weld where too much metal has been deposited. TO CORRECT: shorten the welding time so less metal is deposited into the plug weld hole.

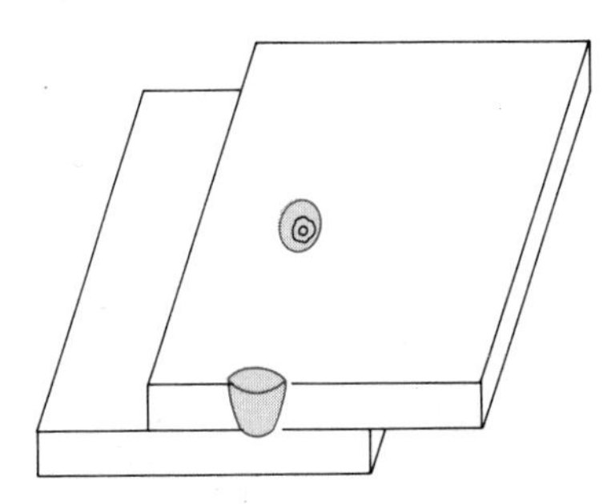

Fig. 11-29. CRATERS exist in the center of the plug weld due to not enough weld metal being placed in the hole during the weld cycle. TO CORRECT: lengthen the time of the weld cycle.

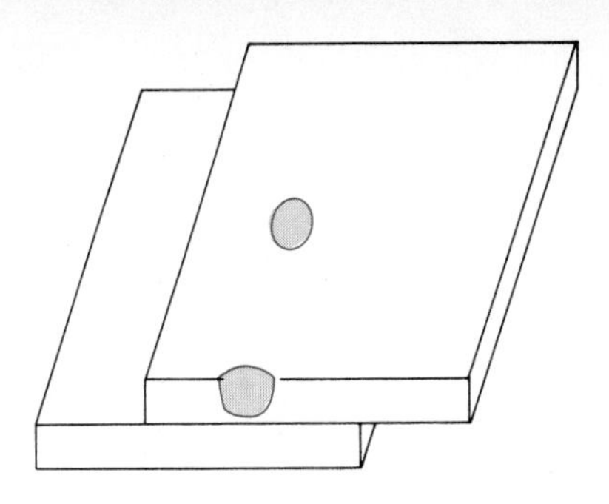

Fig. 11-30. LACK OF PENETRATION is weld metal which did not reach the proper depth in the bottom plate. TO CORRECT: increase amperage, increase weld time, decrease voltage, or decrease stickout.

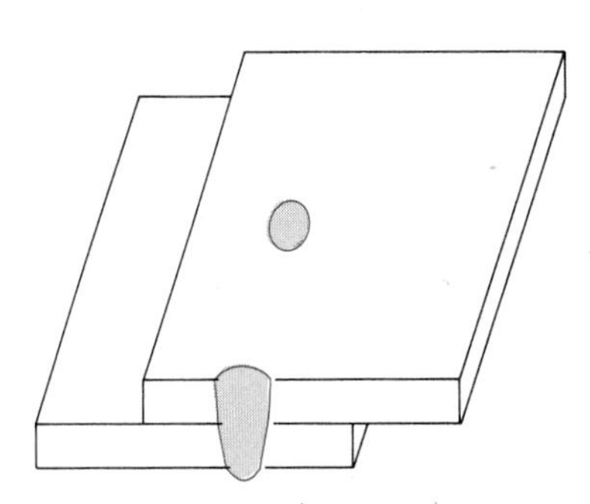

Fig. 11-31. EXCESSIVE PENETRATION is where the welding arc has gone too far into or even burned through the bottom sheet. TO CORRECT: decrease amperage, decrease weld time, increase voltage, or increase stickout.

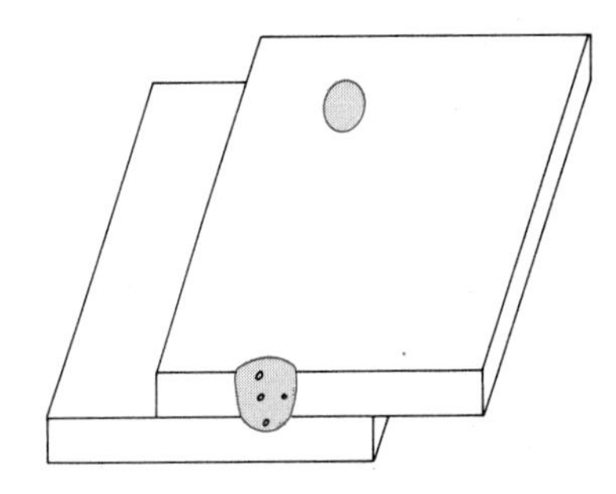

Fig. 11-32. POROSITY occurs in spot welds when the gases formed during heating do not have enough time to rise to the top of the melt and are trapped in the cooling metal. TO CORRECT: increase arc time, increase amperage, decrease voltage, decrease stickout, and make sure that the metal is clean before mating the assembly.

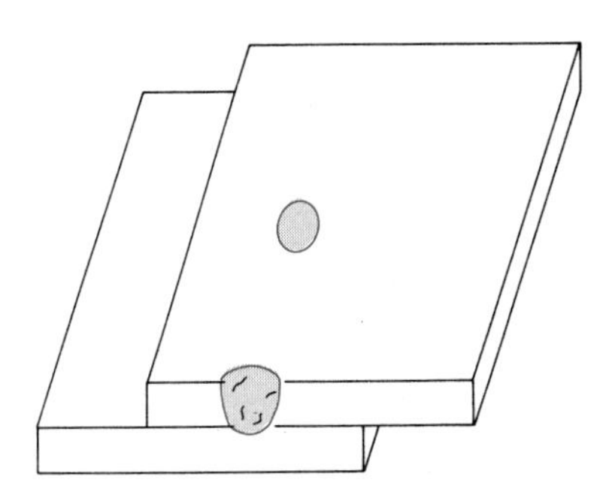

Fig. 11-33. CRACKS happen for the same reasons in spot welds as they do in plug welds. TO CORRECT: use suggestions listed in plug weld cracks.

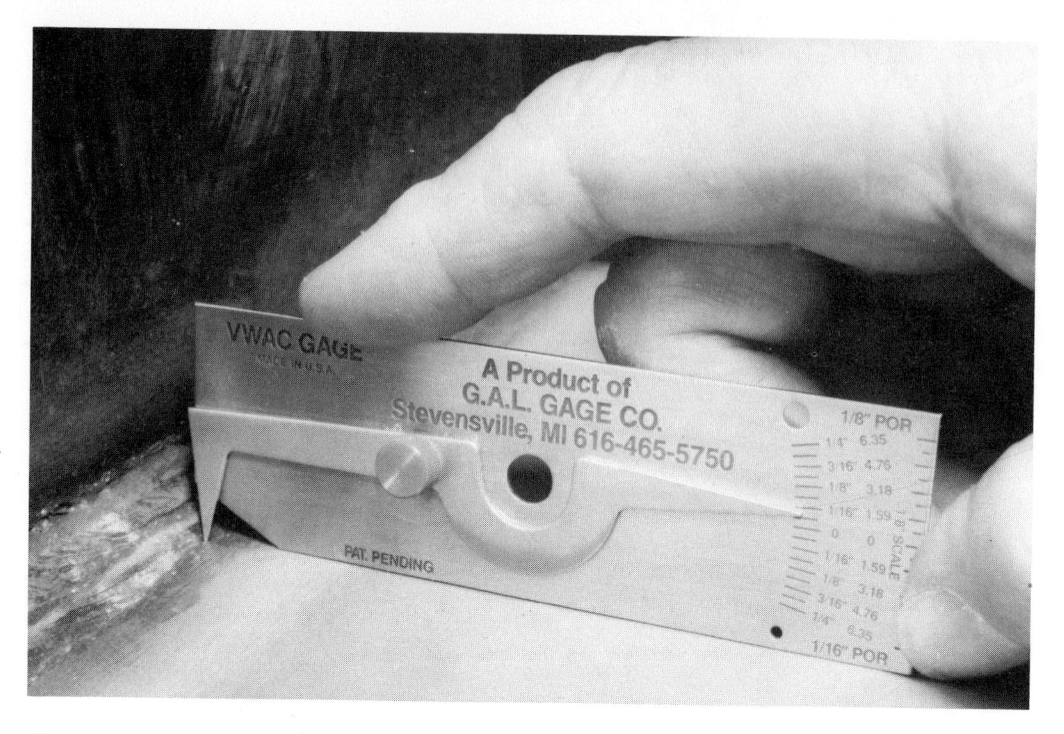

Fig. 11-34. Inspection tool in use to check the undercut at a weld's edge. The amount of undercut is shown on the scale. (G.A.L. Gage Co.)

WELD INSPECTION TOOLS

Various types of inspection tools are used to inspect completed welds for crown height, fillet weld size, undercut, mismatch, and other defects. The welder should become familiar with these tools.

The tool shown in Fig. 11-34 is used to measure undercut, mismatch, and groove weld height.

The tool shown in Fig. 11-35 is used to measure fillet weld leg length size.

The tool shown in Fig. 11-36 is used to measure fillet weld throat size.

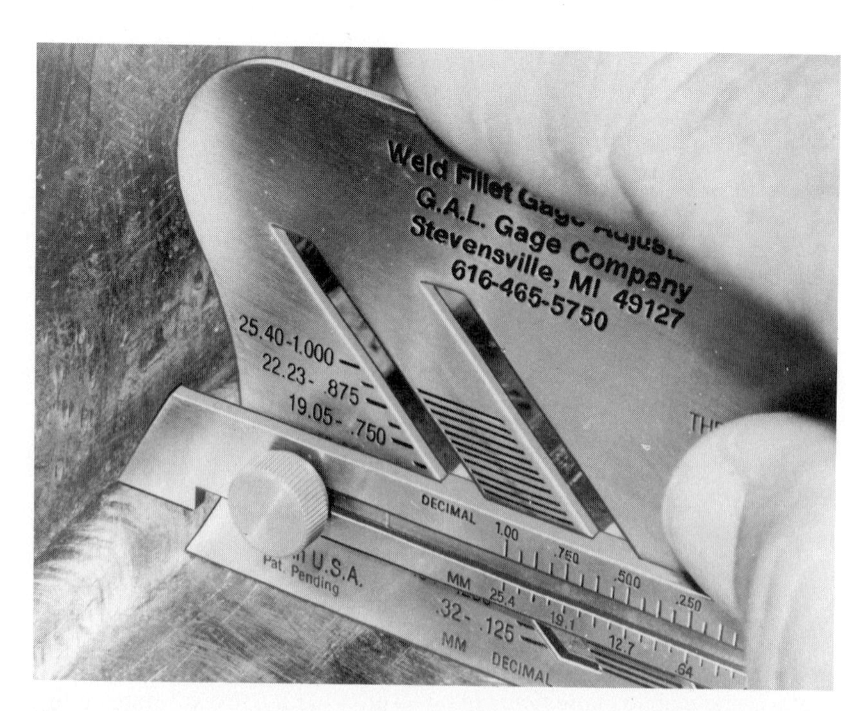

Fig. 11-35. The pointer is placed on the edge of the weld leg and the height of the arrow indicates weld leg size. (G.A.L. Gage Co.)

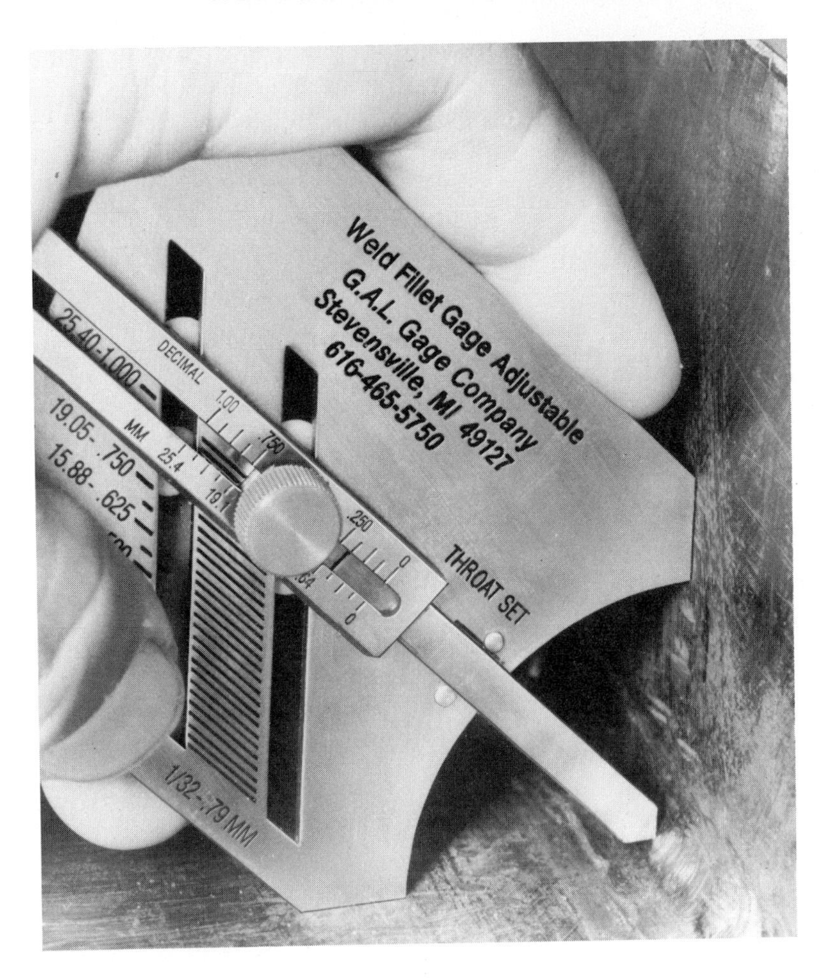

Fig. 11-36. The throat size measurement is used to determine concavity or convexity of the weld. (G.A.L. Gage Co.)

REVIEW QUESTIONS—Chapter 11

1. When the weld does not penetrate completely through a butt weld joint, the condition may be termed lack of penetration. This may also be called __________ __________.
2. When lack of fusion exists between the weld and joint face, it may also be called a __________ __________.
3. When the weld crown flows over the weld edge and does not fuse properly with the edge of the weld joint, it is called an __________.
4. The crown of the weld is peaked in the middle and does not have good contour. This type of weld crown is called a __________ crown.
5. Molten metal shrinks during cooling when the welding stops. A __________ may form if the proper technique or __________ are not used to prevent this defect.
6. Parts which become highly stressed after the metal cools and shrinks may have __________ in the weld or adjacent area.
7. Gas pockets in completed welds are called __________ or __________. If these gas pockets are aligned in a row they are termed __________ __________.
8. Whiskers which form on open butt welds during the root pass are located on the __________ side of the joint.
9. Small pieces of molten metal thrown out of the arc stream during welding attach to the weld area. These drops of metal are called __________.
10. Upon examination of a fillet weld made in the horizontal position, undercut is found. Does this usually exist on the top or the bottom of the weld bead?
11. Fillet welds will generally have linear porosity along the __________ of the joint interface.
12. A lack of penetration exists in plug welds when the weld metal does not extend into the __________ plate.
13. Where do cracks usually occur in plug welds? Why?
14. List at least four defects which are common to plug welds.
15. List four defects which are common to spot welds.

Chapter 12

GENERAL SHOP WELDING PROCEDURES

After studying this chapter, you will be able to:

- Make test welds with various designs, metals, and thicknesses.
- Produce welds using a GMAW schedule.

This chapter details the welding of differing types of steel, stainless steel, and aluminum. The included weld schedules have been made using existing shop procedures. They may vary considerably from the setup charts shown in the reference section of the chapters regarding certain materials. Many factors exist in this welding process. Therefore, it is important that your procedure is confirmed by test welds prior to welding.

A test weld should have the following in common with the required weld: joint design, metal, and material thickness. In other words, copy the weld you are going to make. Upon completion, the test weld should be inspected for quality. The weld must perform its intended function. Testing can be done by visual inspection, cutting welds apart for a macro test, pulling until destruction, bending the welded joint in a press, applying a penetrant, or performing one of many other testing processes.

Once you have started welding, visually check your welds as they are made. The welds should look like your test welds. If they do not, stop and find out why. Minor changes in your schedule may have radical effects on completed welds. These changes are important to the quality of each weld. Tolerances should be established for each parameter or variable.

Good welds can be produced by skilled welders using proven schedules within established tolerances. Faulty welds can be produced by the same welders using unproven schedules or tolerances.

Proven schedules and the weld results are shown in Figs. 12-2 through 12-14.

Jigs are often used to hold parts together during welding. These aluminum frames are being welded with the short-arc mode with argon shielding gas. (ODL, Inc.)

GAS METAL ARC WELDING SCHEDULE

PROCESS MODE *SHORT-ARC*

WELD TYPE *FILLET* POSITION *FLAT/HORIZONTAL*

BASE METAL TYPE *STEEL* BASE METAL THICKNESS *1/16″-1/16″*

WIRE TYPE *E70S-2* WIRE DIA. *.035″* WIRE SPEED (IPM) *190*

GAS TYPE *AR-75* % GAS TYPE *CO_2-25* % FLOW RATE (CFH) *35*

NOZZLE TYPE *STD* NOZZLE DIA. *3/8-.375″*

CONTACT TIP TYPE (long) *X* (short) ____ (standard) ____ DIA. *.035″*

AMPERES *100* VOLTS (arc) *18* STICKOUT *1/4-3/8″*

SPOTWELD TIME (seconds) ____ BURNBACK TIME (seconds) ____

TRAVEL (forehand) ____ (backhand) *X* (vertical up) ____ (down) ____

PRE-WELD CLEANING TYPE *WIRE BRUSH*

Fig. 12-1

GAS METAL ARC WELDING SCHEDULE

PROCESS MODE *SHORT-ARC*

WELD TYPE *EDGE* POSITION *ALL*

BASE METAL TYPE *STEEL* BASE METAL THICKNESS *1/16″*

WIRE TYPE *E70S-6* WIRE DIA. *.030″* WIRE SPEED (IPM) *170*

GAS TYPE *AR-75* % GAS TYPE *CO_2-25* % FLOW RATE (CFH) *35*

NOZZLE TYPE *STD* NOZZLE DIA. *3/8-.375″*

CONTACT TIP TYPE (long) *X* (short) ______ (standard) ______ DIA. *.030″*

AMPERES *85* VOLTS (arc) *17* STICKOUT *3/8″*

SPOTWELD TIME (seconds) ______ BURNBACK TIME (seconds) ______

TRAVEL (forehand) *X* (backhand) ______ (vertical up) ______ (down) *X*

PRE-WELD CLEANING TYPE *WIRE BRUSH*

Fig. 12-2.

GAS METAL ARC WELDING SCHEDULE

PROCESS MODE *SHORT-ARC*

WELD TYPE *FILLET* POSITION *HORIZONTAL*

BASE METAL TYPE *STEEL* BASE METAL THICKNESS *1/8″-1/8″*

WIRE TYPE *E70S-2* WIRE DIA. *.035″* WIRE SPEED (IPM) *280*

GAS TYPE *AR-75* % GAS TYPE *CO_2-25* % FLOW RATE (CFH) *30*

NOZZLE TYPE *STD* NOZZLE DIA. *3/8-.375″*

CONTACT TIP TYPE (long) *X* (short) ____ (standard) ____ DIA. *.035″*

AMPERES *140* VOLTS (arc) *19* STICKOUT *3/8″*

SPOTWELD TIME (seconds) ____ BURNBACK TIME (seconds) ____

TRAVEL (forehand) *X* (backhand) ____ (vertical up) ____ (down) ____

PRE-WELD CLEANING TYPE *WIRE BRUSH*

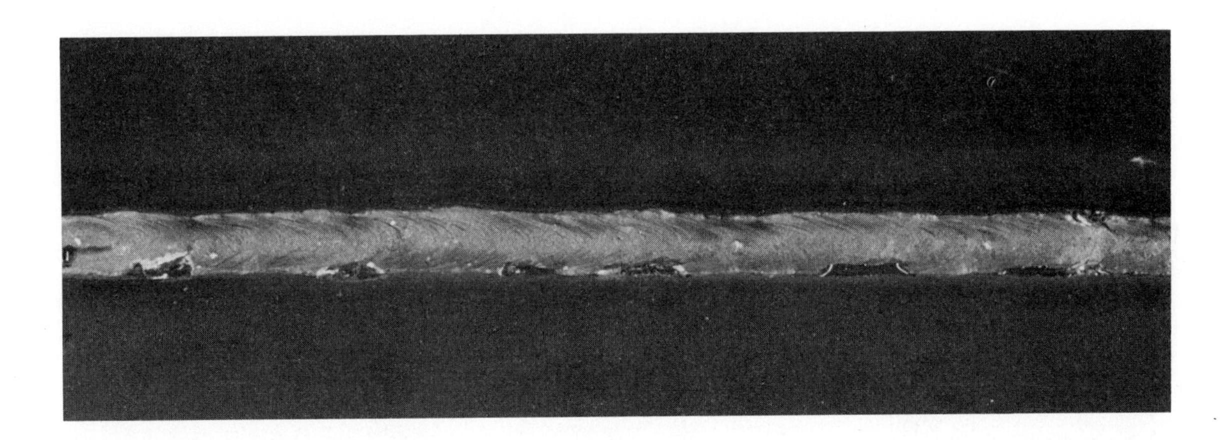

Fig. 12-3.

GAS METAL ARC WELDING SCHEDULE

PROCESS MODE SHORT-ARC

WELD TYPE FILLET/LAP   POSITION FLAT/HORIZONTAL

BASE METAL TYPE STEEL   BASE METAL THICKNESS 1/8″

WIRE TYPE E70S-4   WIRE DIA. .035″   WIRE SPEED (IPM) 290

GAS TYPE ____ %   GAS TYPE CO_2-100 %   FLOW RATE (CFH) 35

NOZZLE TYPE STD   NOZZLE DIA. 3/8-.375″

CONTACT TIP TYPE (long) X   (short) ____   (standard) ____   DIA. .035″

AMPERES 145   VOLTS (arc) 21   STICKOUT 3/8″

SPOTWELD TIME (seconds) ____   BURNBACK TIME (seconds) ____

TRAVEL (forehand) ____   (backhand) X   (vertical up) ____   (down) ____

PRE-WELD CLEANING TYPE WIRE BRUSH

Fig. 12-4.

GAS METAL ARC WELDING SCHEDULE

PROCESS MODE *SHORT-ARC*

WELD TYPE *FILLET* POSITION *HORIZONTAL*

BASE METAL TYPE *STEEL* BASE METAL THICKNESS *3/16″-3/16″*

WIRE TYPE *E70S-2* WIRE DIA. *.035″* WIRE SPEED (IPM) *300*

GAS TYPE *AR-75* % GAS TYPE *CO_2-25* % FLOW RATE (CFH) *35*

NOZZLE TYPE *STD* NOZZLE DIA. *.620-5/8″*

CONTACT TIP TYPE (long) *X* (short) ____ (standard) ____ DIA. *.035″*

AMPERES *165* VOLTS (arc) *20* STICKOUT *3/8″*

SPOTWELD TIME (seconds) ____ BURNBACK TIME (seconds) ____

TRAVEL (forehand) *X* (backhand) ____ (vertical up) ____ (down) ____

PRE-WELD CLEANING TYPE *WIRE BRUSH*

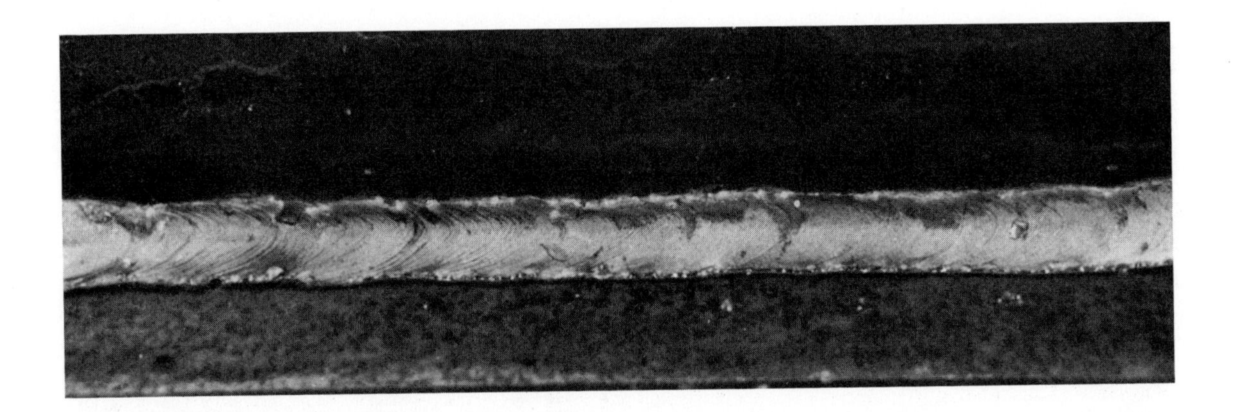

Fig. 12-5.

GAS METAL ARC WELDING SCHEDULE

PROCESS MODE *SHORT-ARC*

WELD TYPE *GROOVE/OPEN* POSITION *FLAT*

BASE METAL TYPE *STEEL* BASE METAL THICKNESS *3/16″*

WIRE TYPE *E70S-2* WIRE DIA. *.035″* WIRE SPEED (IPM) *280*

GAS TYPE *AR-75* % GAS TYPE *CO_2-25* % FLOW RATE (CFH) *30*

NOZZLE TYPE *STD* NOZZLE DIA. *3/8-.375″*

CONTACT TIP TYPE (long) *X* (short) ____ (standard) ____ DIA. *.035″*

AMPERES *140* VOLTS (arc) *19* STICKOUT *3/8″*

SPOTWELD TIME (seconds) ____ BURNBACK TIME (seconds) ____

TRAVEL (forehand) *X* (backhand) ____ (vertical up) ____ (down) ____

PRE-WELD CLEANING TYPE *WIRE BRUSH*

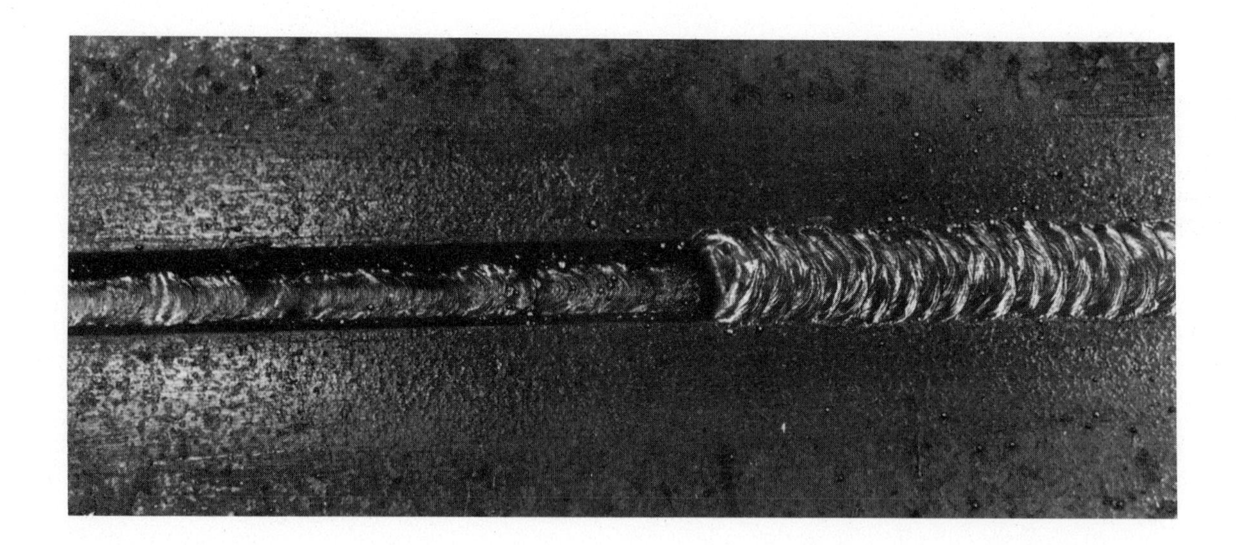

Fig. 12-6.

GAS METAL ARC WELDING SCHEDULE

PROCESS MODE _SPRAY-ARC_

WELD TYPE _FILLET_ POSITION _FLAT_

BASE METAL TYPE _STEEL_ BASE METAL THICKNESS _3/16″-3/16″_

WIRE TYPE _E70S-2_ WIRE DIA. _.035″_ WIRE SPEED (IPM) _375_

GAS TYPE _AR-98_ % GAS TYPE _O_2-2_ % FLOW RATE (CFH) _40_

NOZZLE TYPE _AIR COOLED_ NOZZLE DIA. _.620-5/8″_

CONTACT TIP TYPE (long) ____ (short) _X_ (standard) ____ DIA. _.035″_

AMPERES _200_ VOLTS (arc) _26_ STICKOUT _1/2″_

SPOTWELD TIME (seconds) ________ BURNBACK TIME (seconds) ________

TRAVEL (forehand) _X_ (backhand) ____ (vertical up) ____ (down) ____

PRE-WELD CLEANING TYPE _WIRE BRUSH_

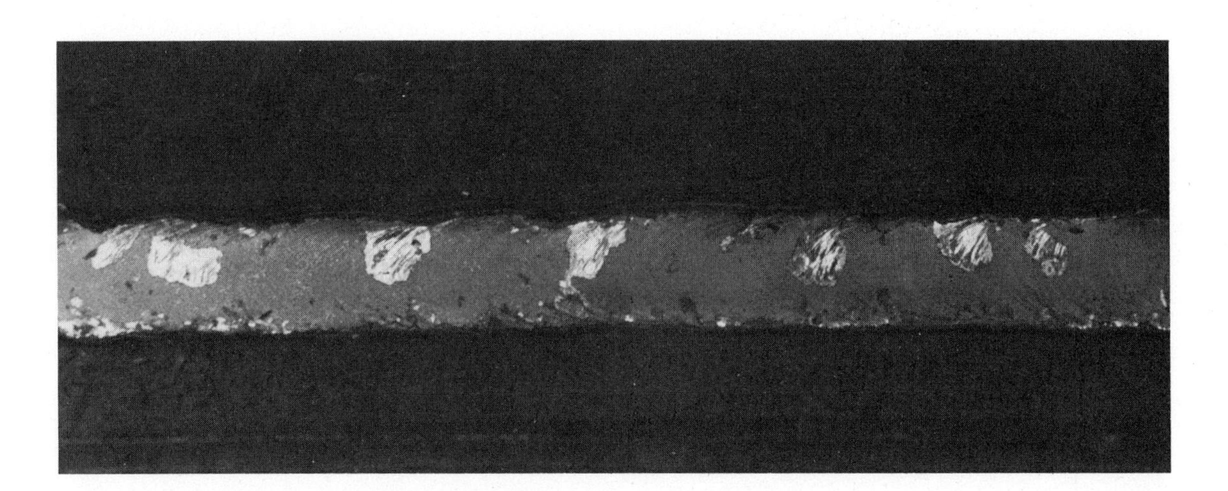

Fig. 12-7.

GAS METAL ARC WELDING SCHEDULE

PROCESS MODE SPRAY-ARC

WELD TYPE GROOVE POSITION FLAT

BASE METAL TYPE STEEL BASE METAL THICKNESS 1/2″-1/2″

WIRE TYPE E70S-2 WIRE DIA. .045″ WIRE SPEED (IPM) 360

GAS TYPE AR-98 % GAS TYPE O_2-2 % FLOW RATE (CFH) 40

NOZZLE TYPE AIR COOLED NOZZLE DIA. 5/8-.620″

CONTACT TIP TYPE (long) ____ (short) X (standard) ____ DIA. .045″

AMPERES 300 VOLTS (arc) 26 STICKOUT 1/2″

SPOTWELD TIME (seconds) ____ BURNBACK TIME (seconds) ____

TRAVEL (forehand) ____ (backhand) X (vertical up) ____ (down) ____

PRE-WELD CLEANING TYPE WIRE BRUSH

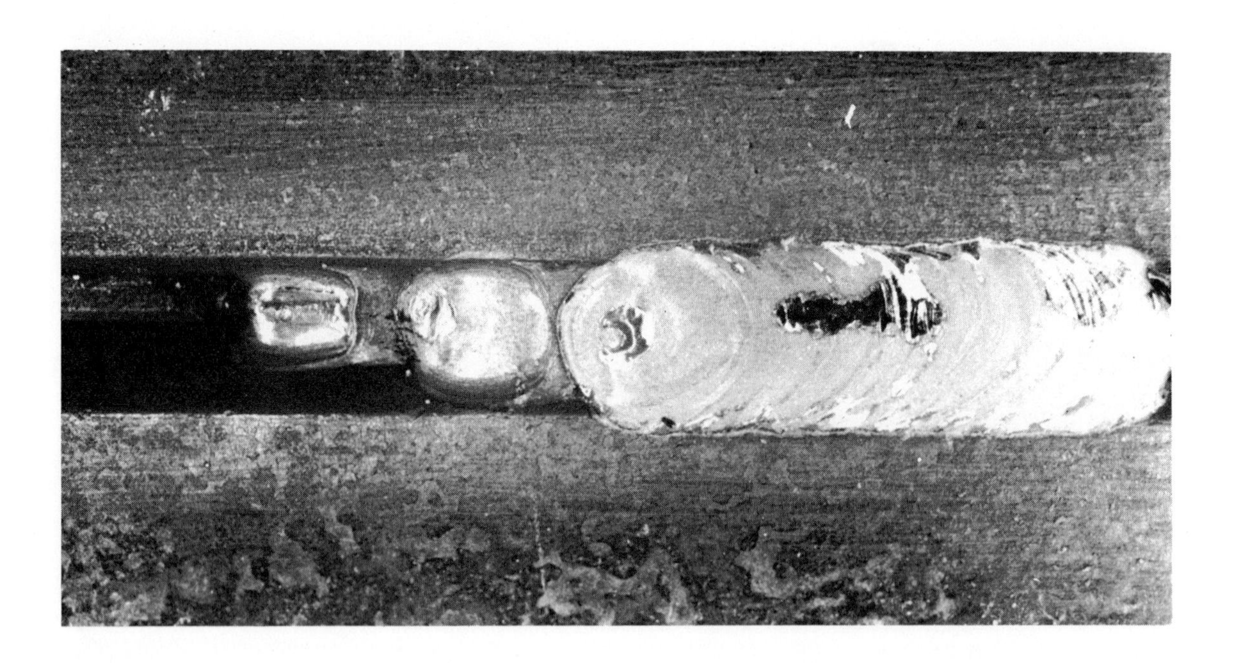

Fig. 12-8.

GAS METAL ARC WELDING SCHEDULE

PROCESS MODE *SPRAY-ARC*

WELD TYPE *PLUG* POSITION *FLAT*

BASE METAL TYPE *STEEL* BASE METAL THICKNESS *3/16″*

WIRE TYPE *E70S-3* WIRE DIA. *.035″* WIRE SPEED (IPM) *370*

GAS TYPE *AR-98* % GAS TYPE *O_2-2* % FLOW RATE (CFH) *40*

NOZZLE TYPE *AIR COOLED* NOZZLE DIA. *.620-.5/8″*

CONTACT TIP TYPE (long) ____ (short) *X* (standard) ____ DIA. *.035″*

AMPERES *190* VOLTS (arc) *25* STICKOUT *1/2″*

SPOTWELD TIME (seconds) ____ BURNBACK TIME (seconds) ____

TRAVEL (forehand) ____ (backhand) ____ (vertical up) ____ (down) ____

PRE-WELD CLEANING TYPE *WIRE BRUSH*

Fig. 12-9.

GAS METAL ARC WELDING SCHEDULE

PROCESS MODE *SHORT-ARC*

WELD TYPE *FILLET* POSITION *HORIZONTAL*

BASE METAL TYPE *ALUM* BASE METAL THICKNESS *.080-.080″*

WIRE TYPE *ER4043* WIRE DIA. *.035″* WIRE SPEED (IPM) *270*

GAS TYPE *ARGON* % GAS TYPE ______ % FLOW RATE (CFH) *30*

NOZZLE TYPE *STD* NOZZLE DIA. *.375-3/8″*

CONTACT TIP TYPE (long) *X* (short) ______ (standard) ______ DIA. *.035″*

AMPERES *60* VOLTS (arc) *15* STICKOUT *3/16″*

SPOTWELD TIME (seconds) ______ BURNBACK TIME (seconds) ______

TRAVEL (forehand) *X* (backhand) ______ (vertical up) ______ (down) ______

PRE-WELD CLEANING TYPE *NONE*

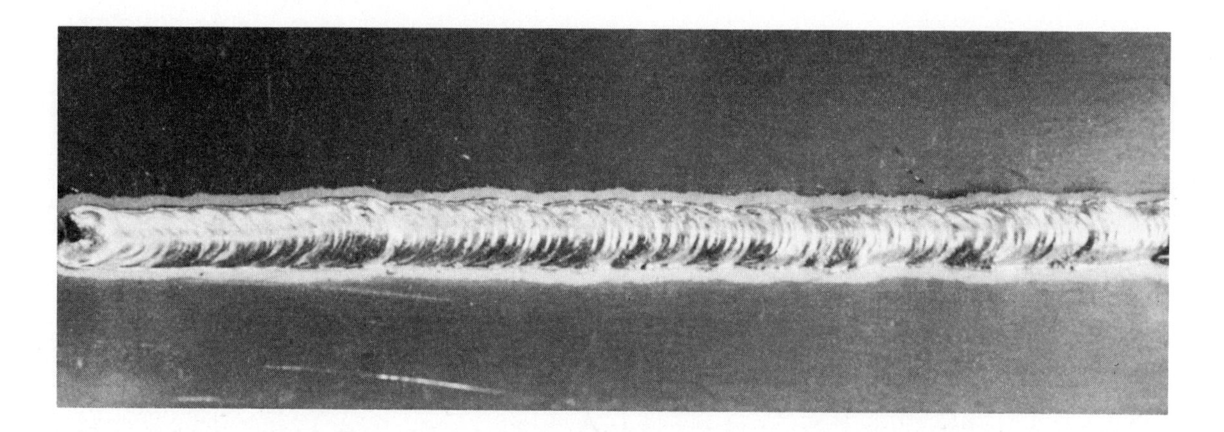

Fig. 12-10.

GAS METAL ARC WELDING SCHEDULE

PROCESS MODE _SPRAY-ARC_

WELD TYPE _FILLET_ POSITION _FLAT_

BASE METAL TYPE _ALUM_ BASE METAL THICKNESS _1/8″-1/8″_

WIRE TYPE _ER4043_ WIRE DIA. _.035″_ WIRE SPEED (IPM) _350″_

GAS TYPE _ARGON_ % GAS TYPE ______ % FLOW RATE (CFH) _35_

NOZZLE TYPE _STD_ NOZZLE DIA. _.500-1/2″_

CONTACT TIP TYPE (long) ____ (short) _X_ (standard) ____ DIA. _.035″_

AMPERES _110_ VOLTS (arc) _21_ STICKOUT _3/8″_

SPOTWELD TIME (seconds) ______ BURNBACK TIME (seconds) ______

TRAVEL (forehand) _X_ (backhand) ____ (vertical up) ____ (down) ____

PRE-WELD CLEANING TYPE _S/STEEL WIRE BRUSH_

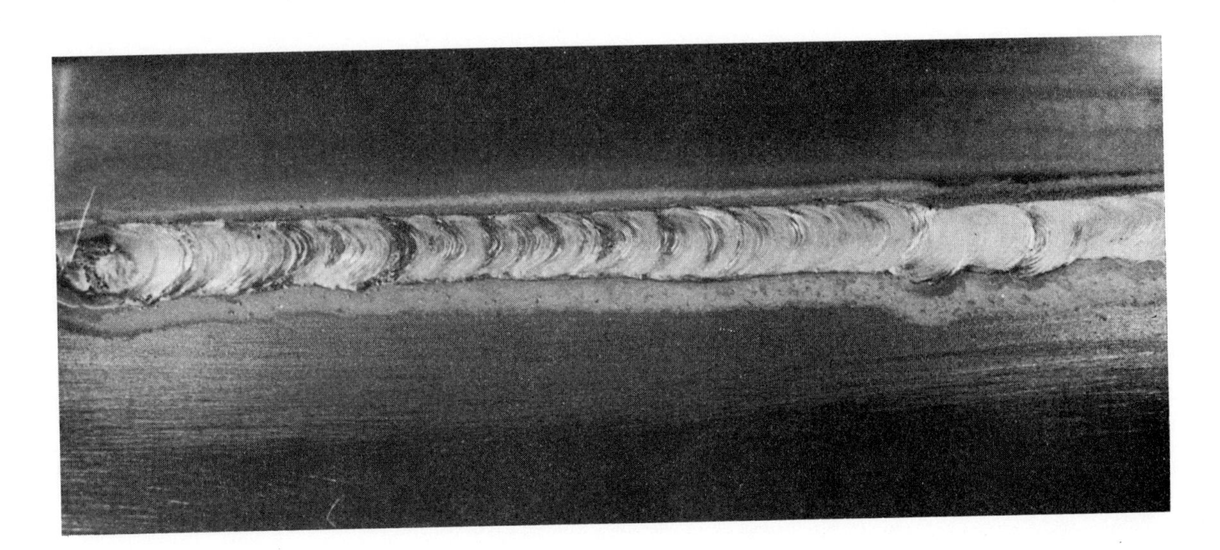

Fig. 12-11.

GAS METAL ARC WELDING SCHEDULE

PROCESS MODE *SPRAY-ARC*

WELD TYPE *FILLET* POSITION *HORIZONTAL*

BASE METAL TYPE *ALUMINUM* BASE METAL THICKNESS *1/4''*

WIRE TYPE *ER5356* WIRE DIA. *3/64''* WIRE SPEED (IPM) *350*

GAS TYPE *ARGON* % GAS TYPE ______ % FLOW RATE (CFH) *40*

NOZZLE TYPE *STD* NOZZLE DIA. *5/8-.620''*

CONTACT TIP TYPE (long) ______ (short) *X* (standard) ______ DIA. *3/64''*

AMPERES *190* VOLTS (arc) *24* STICKOUT *1/2''*

SPOTWELD TIME (seconds) ______ BURNBACK TIME (seconds) ______

TRAVEL (forehand) *X* (backhand) ______ (vertical up) ______ (down) ______

PRE-WELD CLEANING TYPE *DEGREASE – WIRE BRUSH*

Fig. 12-12.

GAS METAL ARC WELDING SCHEDULE

PROCESS MODE *SHORT-ARC*

WELD TYPE *JOGGLE* POSITION *ROTATE-AUTOMATIC*

BASE METAL TYPE *S/STEEL* BASE METAL THICKNESS *14 GA.*

WIRE TYPE *ER316L* WIRE DIA. *.035″* WIRE SPEED (IPM) *275*

GAS TYPE *TRI-MIX* % GAS TYPE ______ % FLOW RATE (CFH) *40*

NOZZLE TYPE *STD* NOZZLE DIA. *.620-5/8″*

CONTACT TIP TYPE (long) *X* (short) ______ (standard) ______ DIA. *.035″*

AMPERES *110* VOLTS (arc) *21* STICKOUT *1/4″*

SPOTWELD TIME (seconds) ______ BURNBACK TIME (seconds) ______

TRAVEL (forehand) *X* (backhand) ______ (vertical up) ______ (down) ______

PRE-WELD CLEANING TYPE *WIRE BRUSH — DEGREASE*

Fig. 12-13.

GAS METAL ARC WELDING SCHEDULE

PROCESS MODE _SPRAY-ARC_

WELD TYPE _FILLET_ POSITION _FLAT_

BASE METAL TYPE _S/STEEL_ BASE METAL THICKNESS _3/16-3/16″_

WIRE TYPE _ER308L_ WIRE DIA. _.035″_ WIRE SPEED (IPM) _410_

GAS TYPE _AR-98_ % GAS TYPE _O_2-2_ % FLOW RATE (CFH) _40_

NOZZLE TYPE _AIR COOLED_ NOZZLE DIA. _.620-5/8″_

CONTACT TIP TYPE (long) ____ (short) _X_ (standard) ____ DIA. _.035″_

AMPERES _170_ VOLTS (arc) _24_ STICKOUT _1/2″_

SPOTWELD TIME (seconds) ____ BURNBACK TIME (seconds) ____

TRAVEL (forehand) _X_ (backhand) ____ (vertical up) ____ (down) ____

PRE-WELD CLEANING TYPE _S/STEEL ETCH_

Fig. 12-14.

Chapter 13

TRUCK TRAILER AND OFF-ROAD VEHICLE WELDING PROCEDURE

After studying this chapter, you will be able to:

- Follow procedures for welding truck trailers and off-road vehicles.
- Inspect and test your own welds.
- Work in compliance with all safety rules.

This chapter details the welding of various joint designs used in making these vehicles. The most common metal used in producing this equipment is steel. It is used in many forms including: pipe, tubing, angle forms, channel forms, beams, plate, forgings, and castings.

Assembly of the component parts is a vital step in constructing a truck trailer. When parts do not fit together, higher stress is placed on the joint which could cause weld failure. Fitup of parts is a major variable in producing reliable welds. It is also essential that all parts are clean before welding. Remove any material which may cause weld defects before starting.

These vehicles are subject to many areas of stress and vibration in use. Therefore, only proven weld procedures should be used. The procedures shown in this chapter may require adapting to your welding system. As stated in Chapter 12, always make a test weld. This helps determine welding parameters and variables for making the best welds. Ultimately, parameters and variables may be concluded through destructive testing, such as break testing. This shows that proper penetration with good fusion but without excessive porosity has been achieved. Quality control procedures must be proven to assure each weld performs its intended function.

The weld schedule shown in Fig. 13-1 specifies the process mode and all welding parameters used in producing a truck trailer. All of the welds shown in Figures 13-2 through Fig. 13-10 were made with this procedure. With these schedules, all welding should be done in the flat or horizontal positions if possible. Whenever welding in the vertical or overhead position, make a test weld first. This helps predict final quality before actual welding.

After welding, visually inspect each joint completely for undercut, overlap, cold shuts, low crowns, fillet weld size, and cracks. If possible, make a dye penetrant test to determine the presence of cracks. When cracks or defects are found, repair these areas before placing the vehicle into service. Even a small undercut, crack, or other defect can cause a weld to fail when stressed during operation.

The weld schedule in Fig. 13-11 specifies the process mode and all of the parameters for constructing the framework of an off-road vehicle. Great stresses are imposed on these vehicles when in use. Therefore, welds must be of a very high grade for the safety of passengers. All framework welds shown in Fig. 13-12 through Fig. 13-15 were made with this procedure. Various components are used on the vehicle for attaching struts, springs, shock absorbers, and other items. The weld procedure used is found in Fig. 13-16. All of the welds in Fig. 13-17 through Fig. 13-20 were made with this procedure.

Vehicle Modification and Repair

During a vehicle's lifetime, changes or repairs are needed to change the structural design or fix broken welds. Replacement parts should fit into place just like the original. Cracked welds, should first be slightly grooved so that the new welds will penetrate into the root of the joint. All grease, oil, or dirt should be removed by wire brushing and the part carefully cleaned with a solvent. Remove rust which may have formed on the metal around the weld area. The weld area should be brought to bright metal condition before welding can begin.

GAS METAL ARC WELDING SCHEDULE

PROCESS MODE *SHORT-ARC*

WELD TYPE *FILLET* POSITION *FLAT/HORIZONTAL*

BASE METAL TYPE *STEEL* BASE METAL THICKNESS *1/8″*

WIRE TYPE *E70S-4* WIRE DIA. *.035″* WIRE SPEED (IPM) *300*

GAS TYPE *AR-75* % GAS TYPE *CO_2-25* % FLOW RATE (CFH) *35*

NOZZLE TYPE *STD* NOZZLE DIA. *1/2″-.500*

CONTACT TIP TYPE (long) *X* (short) ____ (standard) ____ DIA. *.035″*

AMPERES *150* VOLTS (arc) *19* STICKOUT *1/4-3/8″*

SPOTWELD TIME (seconds) ____ BURNBACK TIME (seconds) ____

TRAVEL (forehand) ____ (backhand) *X* (vertical up) ____ (down) ____

PRE-WELD CLEANING TYPE *WIRE BRUSH*

Fig. 13-1. Welding schedule for welding steel truck trailer frame.

Fig. 13-2. Overall view of basic framework.

Fig. 13-3. Channel to channel assembly. Note the reinforcing gusset.

Fig. 13-6. Three piece section channel assembly.

Fig. 13-4. Channel to channel to plate assembly.

Fig. 13-7. Channel to plate assembly.

Fig. 13-5. Channel to channel assembly and a steel rod rope tie.

Fig. 13-8. Angle to angle assembly.

Fig. 13-9. Tubing to tubing to plate assembly.

Fig. 13-10. Tubing to tubing to plate assembly.

GAS METAL ARC WELDING SCHEDULE

PROCESS MODE *SHORT-ARC*

WELD TYPE *FILLET/GROOVE* POSITION *FLAT/HORIZONTAL*

BASE METAL TYPE *STEEL* BASE METAL THICKNESS *1/16 -3/32''*

WIRE TYPE *E70S-4* WIRE DIA. *.030''* WIRE SPEED (IPM) *200*

GAS TYPE *AR-75* % GAS TYPE *CO_2-25* % FLOW RATE (CFH) *35*

NOZZLE TYPE *STD* NOZZLE DIA. *3/8-.375''*

CONTACT TIP TYPE (long) *X* (short) ______ (standard) ______ DIA. *.030''*

AMPERES *105* VOLTS (arc) *19* STICKOUT *1/4-3/8''*

SPOTWELD TIME (seconds) ________ BURNBACK TIME (seconds) ________

TRAVEL (forehand) *X* (backhand) ______ (vertical up) ______ (down) ______

PRE-WELD CLEANING TYPE *WIRE BRUSH*

Fig. 13-11. Welding schedule is for framework tubing assembly.

Fig. 13-12. Typical dune buggy requires many welds.

Fig. 13-13. The end of the tube was cut to fit the outside of the short bearing tube. The cut is often called a "fishmouth" cut.

Fig. 13-14. Tubing and plates are welded together for mounting suspension parts.

Fig. 13-15. Plates are welded together and then assembled to the tube for final welding of the suspension parts.

GAS METAL ARC WELDING SCHEDULE

PROCESS MODE *SHORT-ARC*

WELD TYPE *FILLET* POSITION *FLAT/HORIZONTAL*

BASE METAL TYPE *STEEL* BASE METAL THICKNESS *1/8 -3/16″*

WIRE TYPE *E70S-4* WIRE DIA. *.035″* WIRE SPEED (IPM) *340*

GAS TYPE *AR-75* % GAS TYPE *CO_2-25* % FLOW RATE (CFH) *35*

NOZZLE TYPE *STD* NOZZLE DIA. *1/2″-.500*

CONTACT TIP TYPE (long) *X* (short) ____ (standard) ____ DIA. *.035″*

AMPERES *165* VOLTS (arc) *19* STICKOUT *3/8″*

SPOTWELD TIME (seconds) ____ BURNBACK TIME (seconds) ____

TRAVEL (forehand) ____ (backhand) *X* (vertical up) ____ (down) ____

PRE-WELD CLEANING TYPE *WIRE BRUSH*

Fig. 13-16. Welding schedule for welding plate and tubing suspension parts.

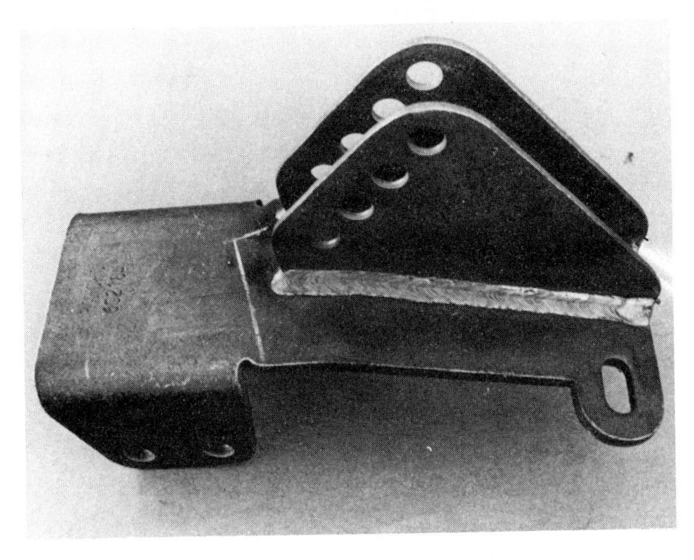

Fig. 13-17. Fillet welds were used to weld these preformed parts in an assembly jig for accurate alignment of the predrilled holes.

Fig. 13-18. Predrilled and preformed plates are attached to the tubing with fillet welds completely around the plate.

When welding a tube with closed ends, always drill a small hole into the tube. This allows hot gases inside the tube to escape. If this is not done, hot gases will expand and exert pressure to escape. Escaping gases will blow away the molten material as you weld. As a result the tube cannot be closed properly.

Fig. 13-19. Suspension assembly of tubing and plates with fillet welds. Note the even contour and shape of the welds.

Fig. 13-20. The fillet weld size (leg length) is equal to the parent metal thickness.

Adding more weld to a joint or rewelding over an old weld does not work on these vehicles. High weld crown, incomplete penetration, and contamination defects cause weld failure when placed under stress. Make sure that the welds you make are correct and will perform their intended job.

Safety

When working on vehicles containing engines, gas tanks, or batteries, always follow the safety rules outlined in Chapter 14. Take the time to do the job right; injuries can be avoided.

Chapter 14

AUTOBODY WELDING PROCEDURES

After studying this chapter, you will be able to:

- Follow safety procedures.
- Prepare an autobody joint for welding.
- Make quality welds using a special autobody welder.
- Remove and replace autobody panels for welding.

Modern automobiles are made using lightweight materials. These materials maintain safety standards without sacrificing strength. This is done by using thinner gauge steels which have been hardened to high tensile strengths before welding. The GMAW process is used with resistance spot welding during manufacturing of many vehicles. For repair of a wrecked vehicle, the GMAW process is widely used. It has proven to be a reliable welding process when applied with proper techniques. GMAW is also used in repairing automobiles that do not use the thin, high tensile steels.

Various filler materials are used by automobile manufacturers during production. However, structural repairs are made with steel filler materials. The steel filler materials include the .023, .030, and .035 in. diameter wires. All can be used with the GMAW process. Minimum heat input is used to avoid overheating high strength base materials. Overheating will temper hardened steel. This reduces its tensile strength.

Safety Procedures

When working on automobiles, remember that you need to protect yourself and others around you by following safety procedures. Welding preparation and actual operation both present many chances for injury. You need to be alert at all times. The following is a list of safety rules.

1. Always wear safety glasses when grinding, drilling, or wire brushing metal.
2. Wear protective clothing when welding.
3. Use proper protective lens in the welding hood. See Fig. 15-1 in reference section.
4. Always disconnect the battery before welding on automobiles.
5. Always cover the battery with wet rags when there is danger that welding sparks may fall on it.
6. Do not weld near open fuel lines.
7. Clean all spilled fuel from area before welding.
8. When welding near a gasoline tank, fill line, or tank vent, make sure all openings are closed. If possible, place wet rags around areas where sparks may fall.
9. Always cover upholstery with damp covers where sparks may fall.
10. Always have a fire extinguisher available.
11. When finished welding, check all areas where sparks or molten metal collect during operation. Be sure that materials, such as insulation, upholstery, or paint have not ignited.

Cleaning Procedures

Many types of sealing materials and insulations are used in making automobiles. These materials must be removed prior to welding. If not removed beforehand, they might burn. This can contaminate the environment for both the weld and the welder.

In addition to being toxic, newly formed gases can cause several weld defects:

1. Porosity.
2. Lack of fusion on the weld boundary.
3. Cracks in the weld metal.

Dirt, rust, scale, and oil will also cause the same problems in welds. Therefore, these areas MUST be cleaned before welding. Disc sanders, files, grinders, and power wire brushes may be used to remove these materials. ALWAYS WEAR SAFETY SHIELDS WHEN PERFORMING THIS OPERATION. A clean area ready for welding is shown in Fig. 14-1.

Zinc coating which has been applied to prevent corrosion does not have to be removed before welding. Zinc has a lower melting point and rises in the molten pool having no effect on weld strength.

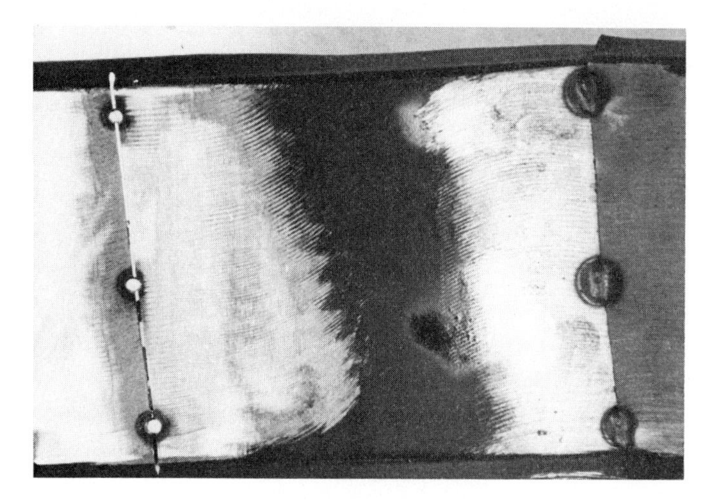

Fig. 14-1. The painted area has been cleaned with a polishing disc and tackwelded before welding the seam.

Equipment

The welding industry has developed special welding units for this type of work. They have controls for welding light gauge steels and operate on either 110 VAC or 208 VAC power. A 110 VAC unit is shown in Fig. 14-2 and a 208 VAC unit in Fig. 14-3. These machines usually operate under adverse conditions common in repair shop use. Therefore, they require special maintenance procedures for proper operation. It is strongly advised that the operating manual maintenance procedures be followed to produce the best results from the machine.

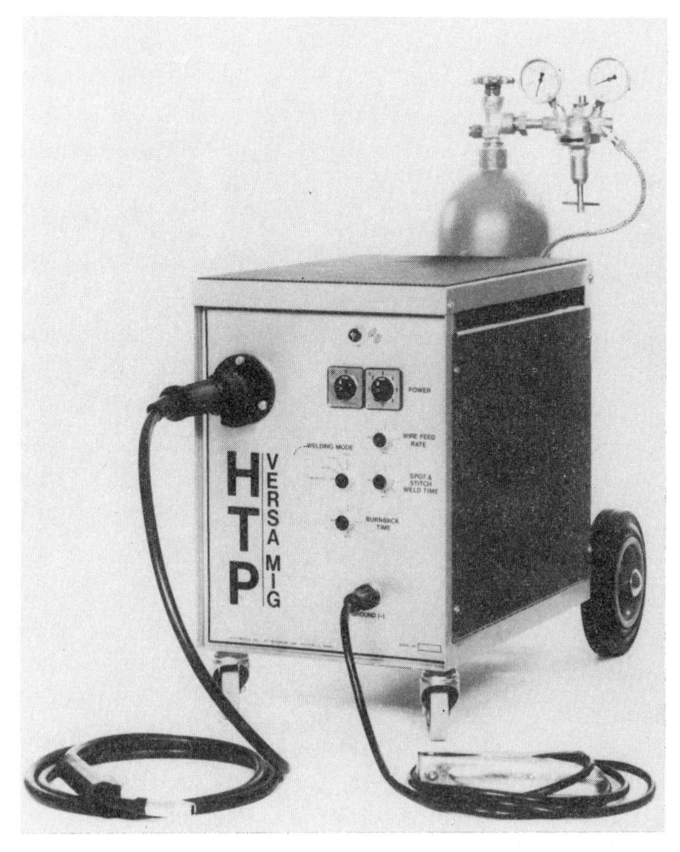

Fig. 14-3. 208 VAC autobody welding unit. (HTP America)

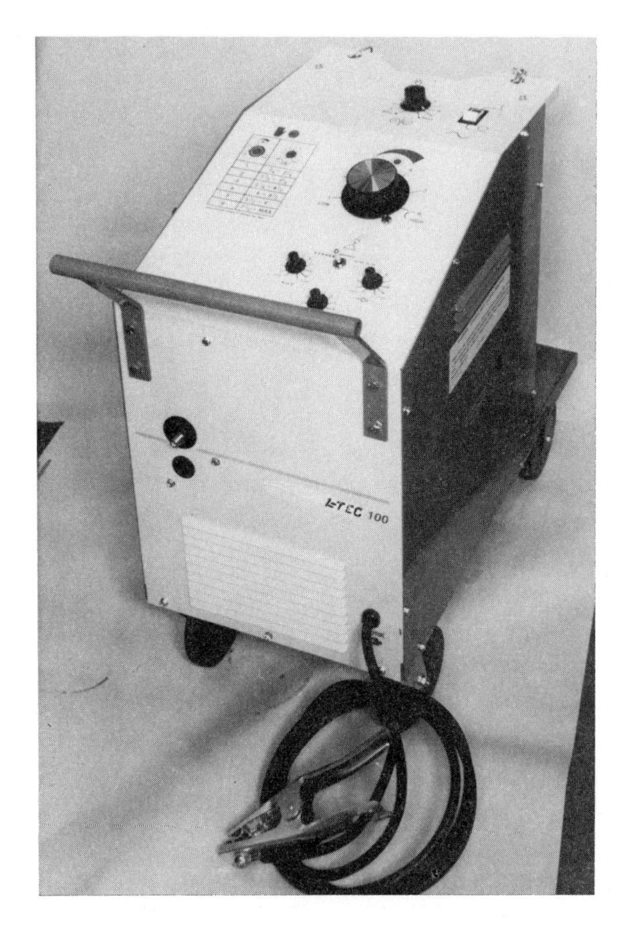

Fig. 14-2. 110 VAC autobody welding unit. (L Tech)

Special Tools

Tools are used in the automotive industry to prepare for lap joints where plug and joggle seam welds are made. A hole puncher is used to make either 1/4 or 5/16 in. diameter holes for plug welds. The puncher may either be a manual or pneumatic unit. Look at Fig. 14-4 and 14-5. Joggle joints, Fig. 14-6, are prepared by bending one piece of the assembly with a tool like in Fig. 14-7. This process allows pieces of abutting metal to mate on the outside of parts in a flat plane for welding seam joints.

Welding Procedures

Welding procedure depends on many factors:

1. Type of joint.
2. Type of weld.
3. Thickness of material.
4. Size of filler material.
5. Type of shielding gas.

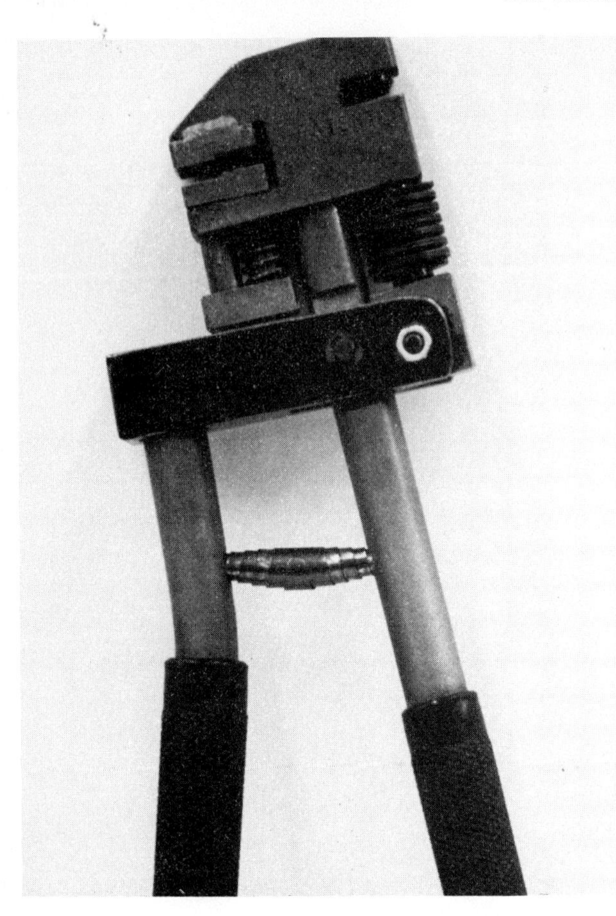

Fig. 14-4. This manual tool is used to punch holes and make flanges for joggle joints.

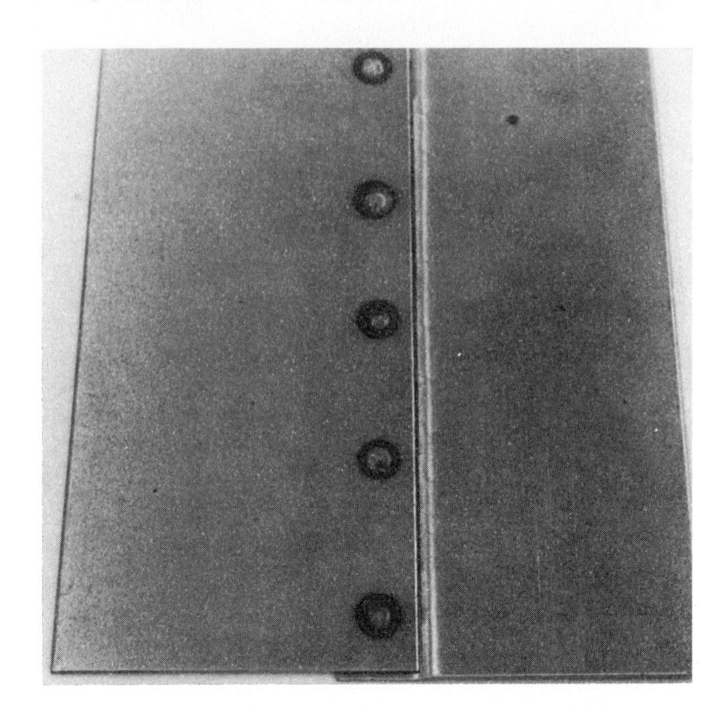

Fig. 14-6. Joggle joint. The small discolored areas are resistance spot welds used to hold the assembly together. GMAW tack welds and clamps may also be used.

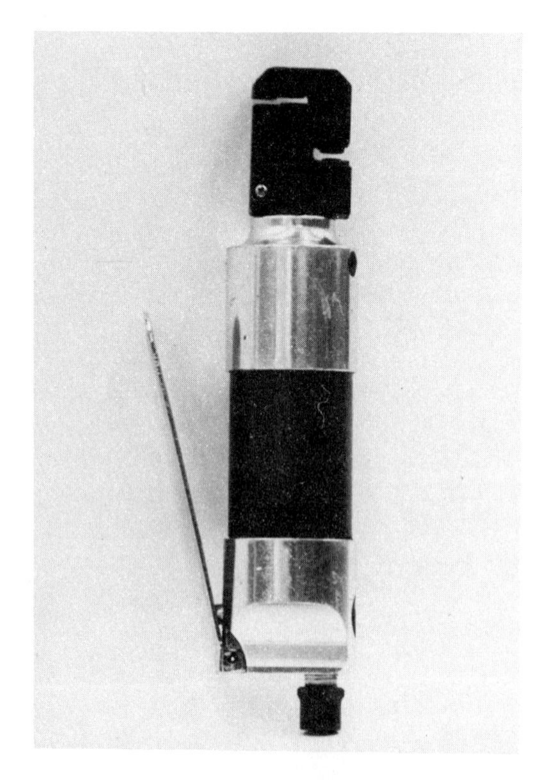

Fig. 14-5. Where many holes are required, a pneumatic hole puncher is used.

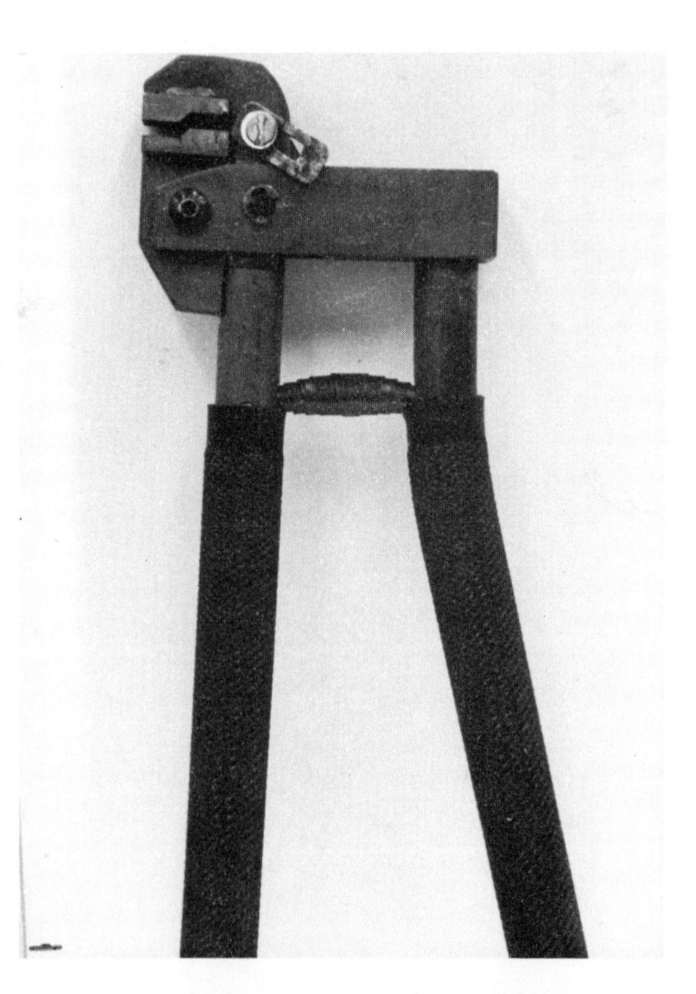

Fig. 14-7. Joggle joint flange tool.

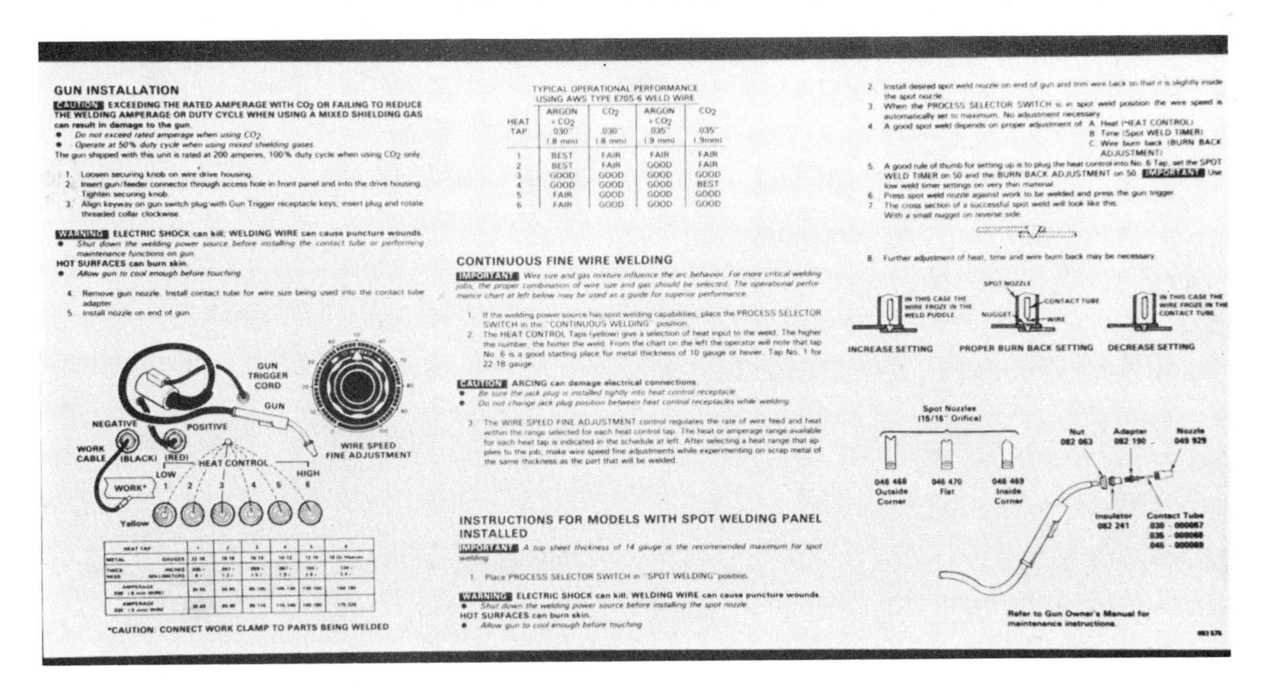

Fig. 14-8. Weld schedule on door of welding machine. (Miller Electric Co.)

Most autobody welding equipment does not have voltmeters or ammeters for setting machine welding parameters. They have established procedures based on the aforementioned factors. These procedures are included on or within the welding machine. They are also included in each machine's operating manual. Remember, these parameters are general conditions. Test welds should be made to determine the machine setup for quality welds. Fig. 14-8 shows a schedule placed on the machine door by the manufacturer. Welding a test sample, as in Fig. 14-9, confirms that the parameters are right for making the required weld.

Figs. 14-10 through 14-14 are actual shop procedures for different welds common in this industry. Amperage and voltage are not shown because these meters do not exist on autobody welding machines.

To follow these procedures, use the wire feed (inches per minute), and adjust the heat rheostat (voltage) until the arc gap is very close with minimum wire stickout.

Fig. 14-9. This test weldment simulates the material thickness and joint design of the desired weld. Visual and destructive test may be performed to confirm weld quality.

GAS METAL ARC WELDING SCHEDULE

PROCESS MODE *SHORT-ARC*

WELD TYPE *PLUG* POSITION *FLAT*

BASE METAL TYPE *STEEL* BASE METAL THICKNESS *.030 TOP*

WIRE TYPE *E70S-6* WIRE DIA. *.023″* WIRE SPEED (IPM) *140*

GAS TYPE *AR-75* % GAS TYPE *CO_2-25* % FLOW RATE (CFH) *35*

NOZZLE TYPE *STD* NOZZLE DIA. *.500-1/2″*

CONTACT TIP TYPE (long) *X* (short) ____ (standard) ____ DIA. *.023″*

AMPERES ________ VOLTS (arc) ________ STICKOUT *1/4″*

SPOTWELD TIME (seconds) ________ BURNBACK TIME (seconds) ________

TRAVEL (forehand) ____ (backhand) ____ (vertical up) ____ (down) ____

PRE-WELD CLEANING TYPE *DEGREASE*

Fig. 14-10. Lap joint plug weld. The upper part of the figure shows the crown of the weld. The lower part of the figure shows the penetration of the weld. The penetration through the lower plate shows an acceptable weld.

GAS METAL ARC WELDING SCHEDULE

PROCESS MODE SHORT-ARC

WELD TYPE FILLET   POSITION FLAT

BASE METAL TYPE STEEL   BASE METAL THICKNESS .035″ TOP

WIRE TYPE E70S-6   WIRE DIA. .023″   WIRE SPEED (IPM) 168″

GAS TYPE AR-75 %   GAS TYPE CO_2-25 %   FLOW RATE (CFH) 30

NOZZLE TYPE STD   NOZZLE DIA. .500-1/2″

CONTACT TIP TYPE (long) X (short) ____ (standard) ____ DIA. .023″

AMPERES ________   VOLTS (arc) ________   STICKOUT 1/4″

SPOTWELD TIME (seconds) ________   BURNBACK TIME (seconds) ________

TRAVEL (forehand) X (backhand) ____ (vertical up) ____ (down) ____

PRE-WELD CLEANING TYPE DEGREASE

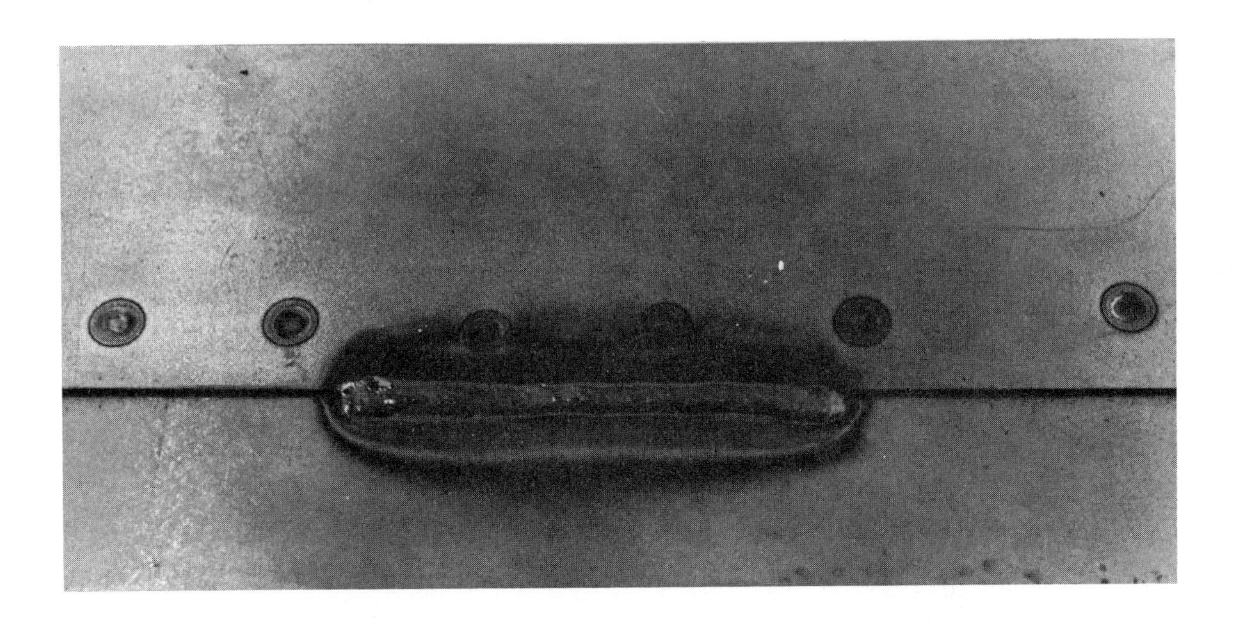

Fig. 14-11. Lap joint fillet weld.

GAS METAL ARC WELDING SCHEDULE

PROCESS MODE *SHORT-ARC*

WELD TYPE *SPOT* POSITION *HORIZONTAL*

BASE METAL TYPE *STEEL* BASE METAL THICKNESS *.059″*

WIRE TYPE *E70S-6* WIRE DIA. *.030″* WIRE SPEED (IPM) *360″*

GAS TYPE *AR-75* % GAS TYPE *CO_2-25* % FLOW RATE (CFH) *30*

NOZZLE TYPE *SPOT* NOZZLE DIA. *.620-5/8″*

CONTACT TIP TYPE (long) ______ (short) *X* (standard) ______ DIA. ______

AMPERES ____________ VOLTS (arc) ____________ STICKOUT *1/8″*

SPOTWELD TIME (seconds) *2* BURNBACK TIME (seconds) *1/4*

TRAVEL (forehand) ______ (backhand) ______ (vertical up) ______ (down) ______

PRE-WELD CLEANING TYPE *SAND-GRIND-DEGREASE*

Fig. 14-12. Lap joint spot weld.

GAS METAL ARC WELDING SCHEDULE

PROCESS MODE SHORT-ARC

WELD TYPE BUTT POSITION FLAT

BASE METAL TYPE STEEL BASE METAL THICKNESS .030″

WIRE TYPE E70S-4 WIRE DIA. .023″ WIRE SPEED (IPM) 160″

GAS TYPE AR-75 % GAS TYPE CO_2-25 % FLOW RATE (CFH) 30

NOZZLE TYPE STD NOZZLE DIA. .375-3/8″

CONTACT TIP TYPE (long) X (short) ____ (standard) ____ DIA. .023″

AMPERES ________ VOLTS (arc) ________ STICKOUT 1/4″

SPOTWELD TIME (seconds) ________ BURNBACK TIME (seconds) ________

TRAVEL (forehand) ____ (backhand) ____ (vertical up) ____ (down) ____

PRE-WELD CLEANING TYPE GRIND-SAND-WIRE BRUSH

Fig. 14-13. Butt joint tackweld and seamweld. The grinder is used to remove the weld crown on tackwelds before making the longseam weld. This operation assists penetration throughout the joint.

GAS METAL ARC WELDING SCHEDULE

PROCESS MODE *SHORT-ARC*

WELD TYPE *JOGGLE* POSITION *FLAT*

BASE METAL TYPE *STEEL* BASE METAL THICKNESS *.024″/.030″*

WIRE TYPE *E70S-4* WIRE DIA. *.023″* WIRE SPEED (IPM) *180*

GAS TYPE *AR-75* % GAS TYPE *CO_2-25* % FLOW RATE (CFH) *30*

NOZZLE TYPE *STD* NOZZLE DIA. *.500-1/2″*

CONTACT TIP TYPE (long) *X* (short) ____ (standard) ____ DIA. *.023″*

AMPERES ____ VOLTS (arc) ____ STICKOUT *3/8″*

SPOTWELD TIME (seconds) ____ BURNBACK TIME (seconds) ____

TRAVEL (forehand) *X* (backhand) ____ (vertical up) ____ (down) ____

PRE-WELD CLEANING TYPE *DEGREASE*

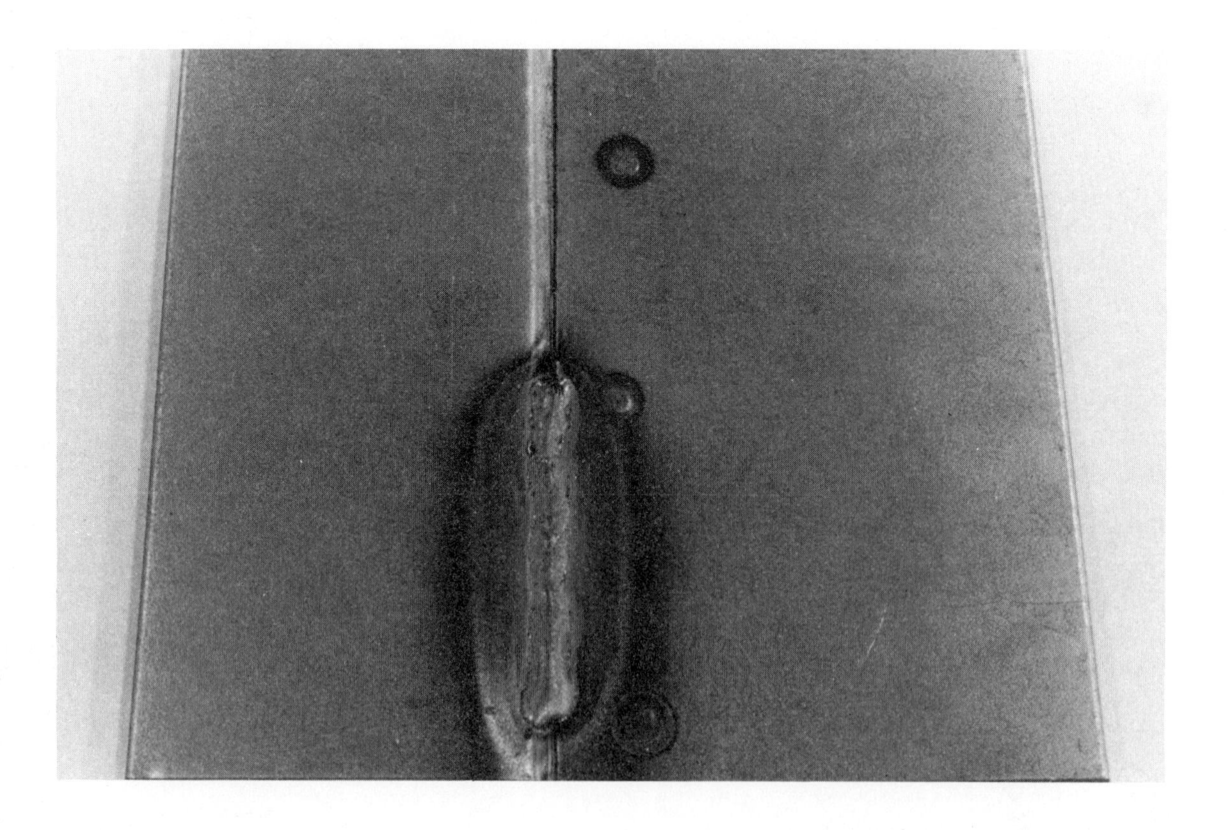

Fig. 14-14. Joggle joint seam weld.

Unibody Repair Procedures

Many modern automobiles are assembled with resistance spot welds made by robots. These spot welds must be removed to take off damaged panels. This is done by using a drill ground to a flat angle and routers. A series of routers are pictured in Fig. 14-15. Where additional welds are desired in the replacement panel, holes are punched or drilled. The manufacturer's recommendations for hole spacing on replacement parts should be followed for proper strength in the joint area. Where this dimension is unknown, one inch to 1 1/2 in. spacing can be used. The joint area is then cleaned and the new components assembled. Fig. 14-16 and Fig. 14-17 show new replacement panels being held in place by clamps prior to welding. Where clamps cannot be used, drill small holes in the panels and use sheet metal screws for holding the panels together. After tackwelding the joint, remove the screws. Then drill out the upper panel screw hole for a plug weld.

To assure good starting of the arc, you may cut the end of the electrode to the desired stickout with a wire cutter, Fig. 14-18. This is especially helpful when making plug and spot welds where any additional wire stickout might cause cold shuts or lack of fusion.

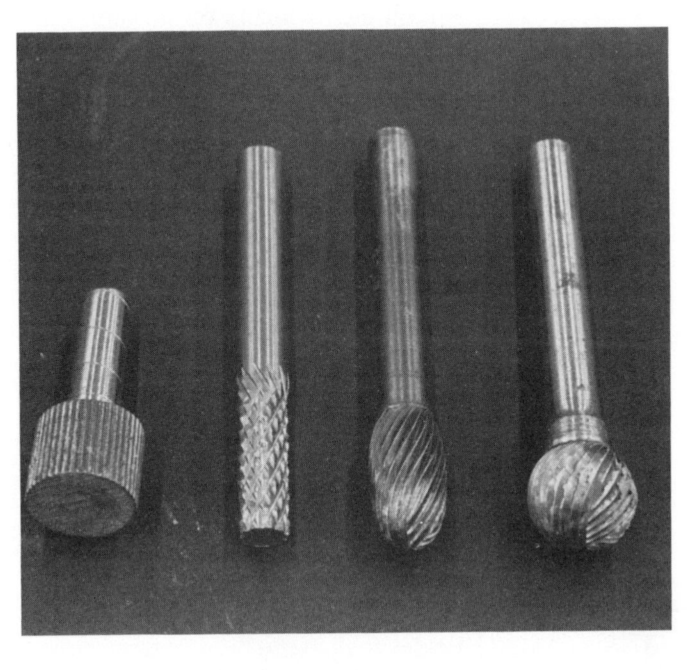

Fig. 14-15. Various types of routers are used to remove original spot welds.

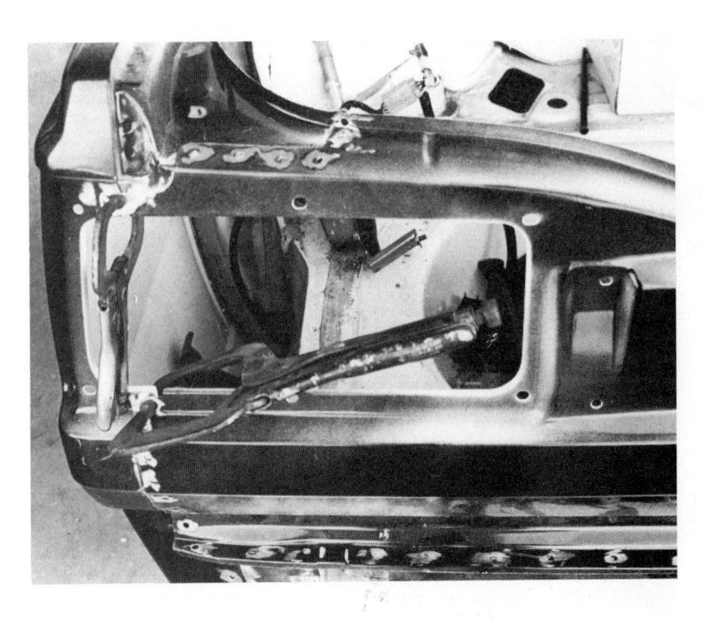

Fig. 14-17. Replacement panel is aligned and clamped prior to welding.

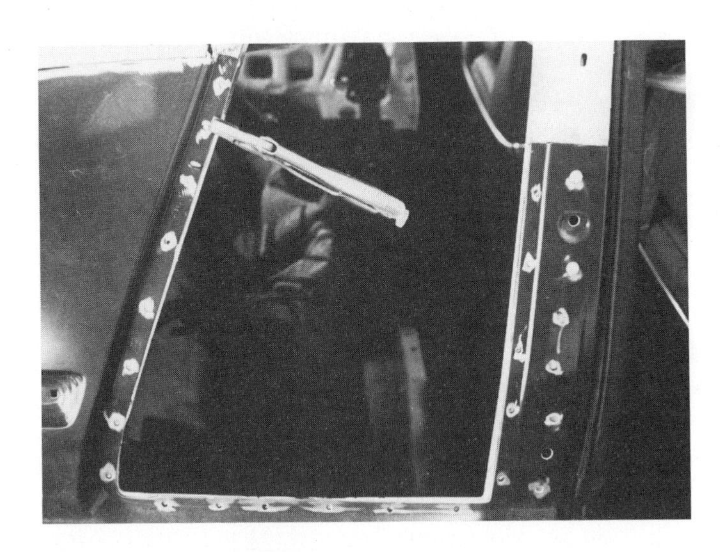

Fig. 14-16. All cleaning is done before assembly of replacement panel.

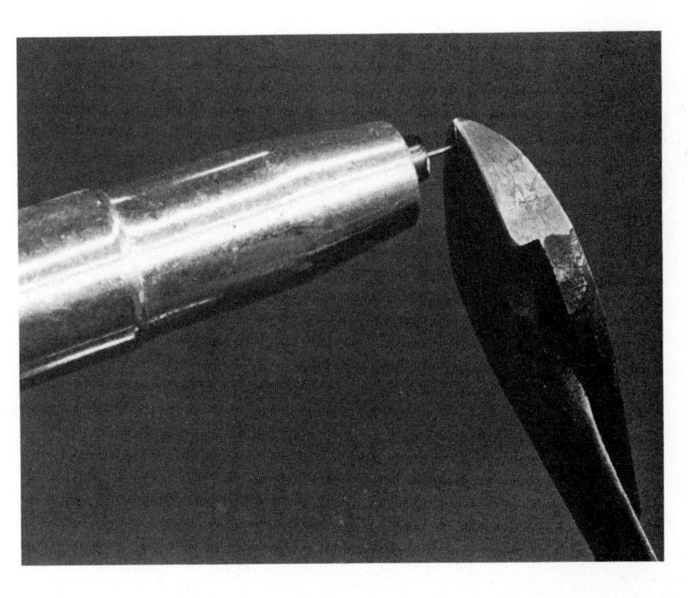

Fig. 14-18. Remove oxidized wire and excess stickout before welding.

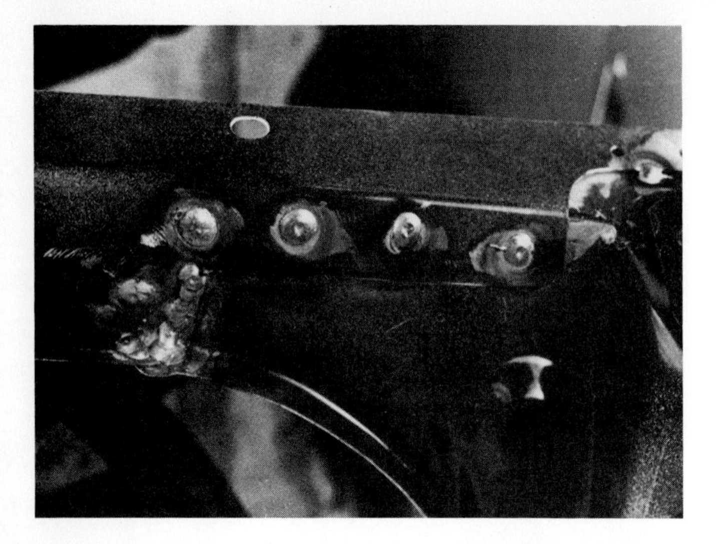

Fig. 14-19. These welds have a closer spacing than the original factory spot welds.

Fig. 14-20. The welds have been completed. They are now ready for grinding and polishing prior to painting.

Fig. 14-19 and Fig. 14-20 show butt, fillet, plug, and spot welds that complete the repair of this vehicle.

REVIEW QUESTIONS—Chapter 14

1. Compare modern steels used by the automotive industry with the steel they have used in the past.
2. Aside from GMAW, what other welding process is commonly used in the automobile industry?
3. Using GMAW for vehicle repair requires approved techniques for ________ welds.
4. What diameter wires are used for welding unibody automobiles?
5. When too much ________ is used during the welding of high strength steels, the material loses its tensile strength.
6. Improper removal of insulation, paint, and dirt from the weld joint before welding results in ________. Two other problems including ________ and ________ may also occur.
7. Does the zinc coating which has been applied for corrosion protection have to be removed before welding?
8. Does zinc have a higher or lower melting point than steel?
9. Why do welding machines used in the autobody repair area require special maintenance?
10. Plug weld hole sizes are either ________ or ________ inch diameter.
11. Joggle joints are made by ________ one piece of the assembly. This allows the top side of the joint to be made in a ________ plane.
12. Hole punches are either manual or ________.
13. What are the main adjustments on an autobody welding machine?
14. What should always be done before beginning any procedure with the GMAW process?
15. The removal of damaged panels from the vehicle is done with ________ ________ drills and routers.
16. Cutting the end of the wire to a predetermined stickout assures ________ ________ of the arc. When this is omitted, the weld may have ________ ________ or ________ ______ ________.

SAFETY TEST

1. What common precaution should be taken when welding, grinding, drilling, or using a power wire brush?
2. How does a smart welder protect his body from arc rays, sparks, and hot metal while welding?
3. How does a welder make sure his eyes are properly protected from arc radiation?
4. What step should always be taken before beginning to weld on automobiles with any arc welding process?
5. Always ________ the battery with wet rags, when sparks may fall.
6. Never weld near an open ________ line.
7. Always clean any spilled ________ from the area before welding.
8. When welding near gasoline tanks, vents, or fill lines, always make sure the openings are ________ and covered with wet rags.
9. When welding near upholstered material, cover it with ________ ________ where sparks may fall.
10. Always have a fire ________ available before welding starts.
11. After welding is completed, check ________ ________ where sparks may have collected so that a fire does not start.
12. You are responsible for working in a safe environment for both yourself and others near you. Keep alert, think ________, and you won't get hurt.

Chapter 15

GMAW REFERENCE SECTION

FILTER PLATE LENSES

WELDING PROCESS	Approximate welding range (in amps)	Federal Specification filter shade required
Metal-arc welding (coated electrodes) Continuous covered electrode welding Carbon Dioxide shield continuous covered electrode welding	100	8 or 9
	100-300	10 or 11
	Over 300	12 or 14
Metal-arc welding (bare wire) Carbon-arc welding Inert-gas metal-arc welding Atomic hydrogen welding	200	10 or 11
	Over 200	12 or 14
Automatic carbon dioxide shield metal-arc welding (bare wire)	Over 500	15 or 16
Inert-gas tungsten-arc welding	15	8
	15-75	9
	75-100	10
	100-200	11
	200-250	12
	250-300	14

Fig. 15-1. Filter plate lenses are designed to protect welders against harmful infrared and ultraviolet rays. The lenses are heat treated and manufactured to meet or exceed ANSI Z-87.1 and Federal Specification GGG-H-211C. Heat-treated lenses are identified by the letter "H" following the shade number. (Thermacote-Welco Company)

Gas Metal Arc Welding Handbook

AMERICAN WELDING SOCIETY

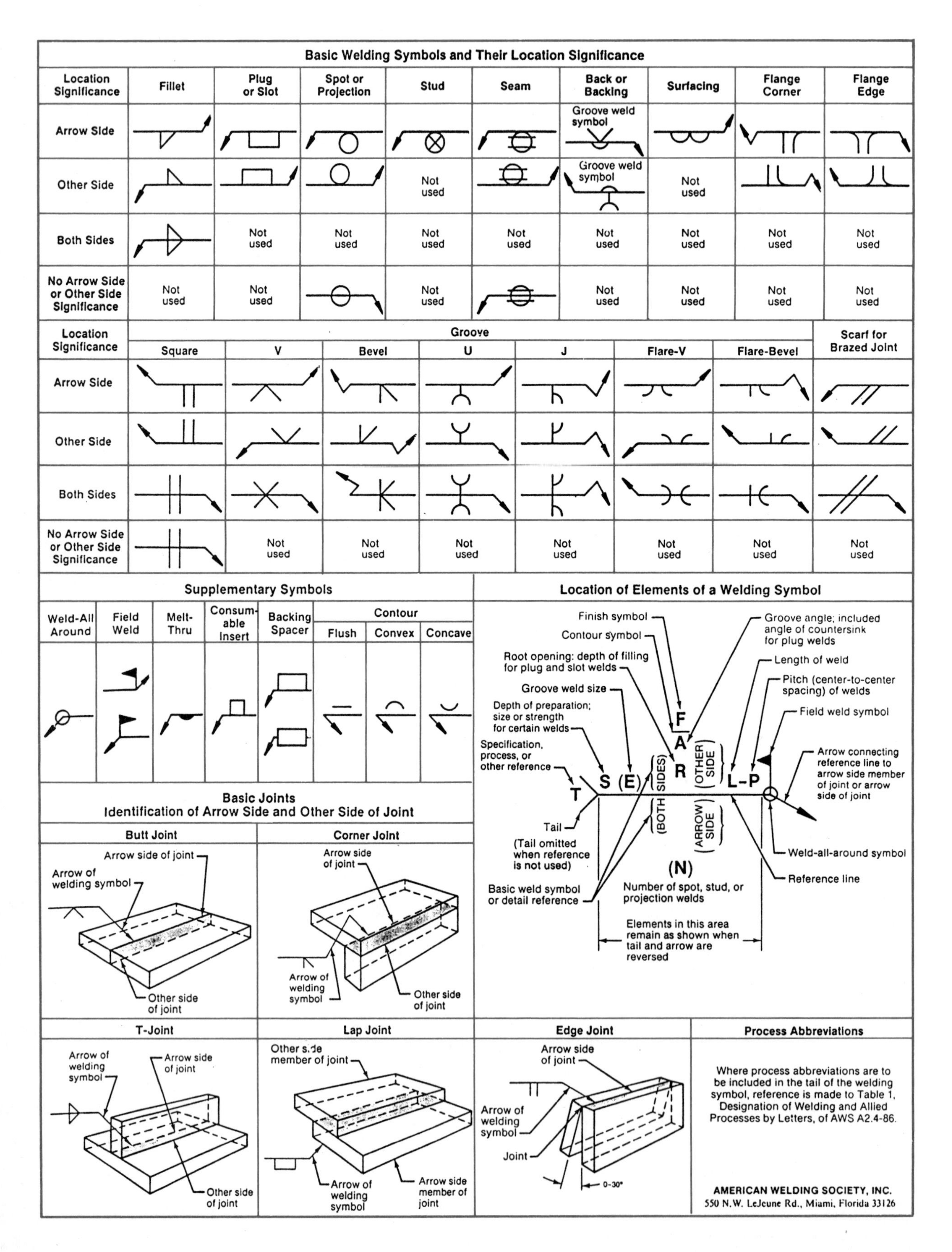

Fig. 15-2.

STANDARD WELDING SYMBOLS

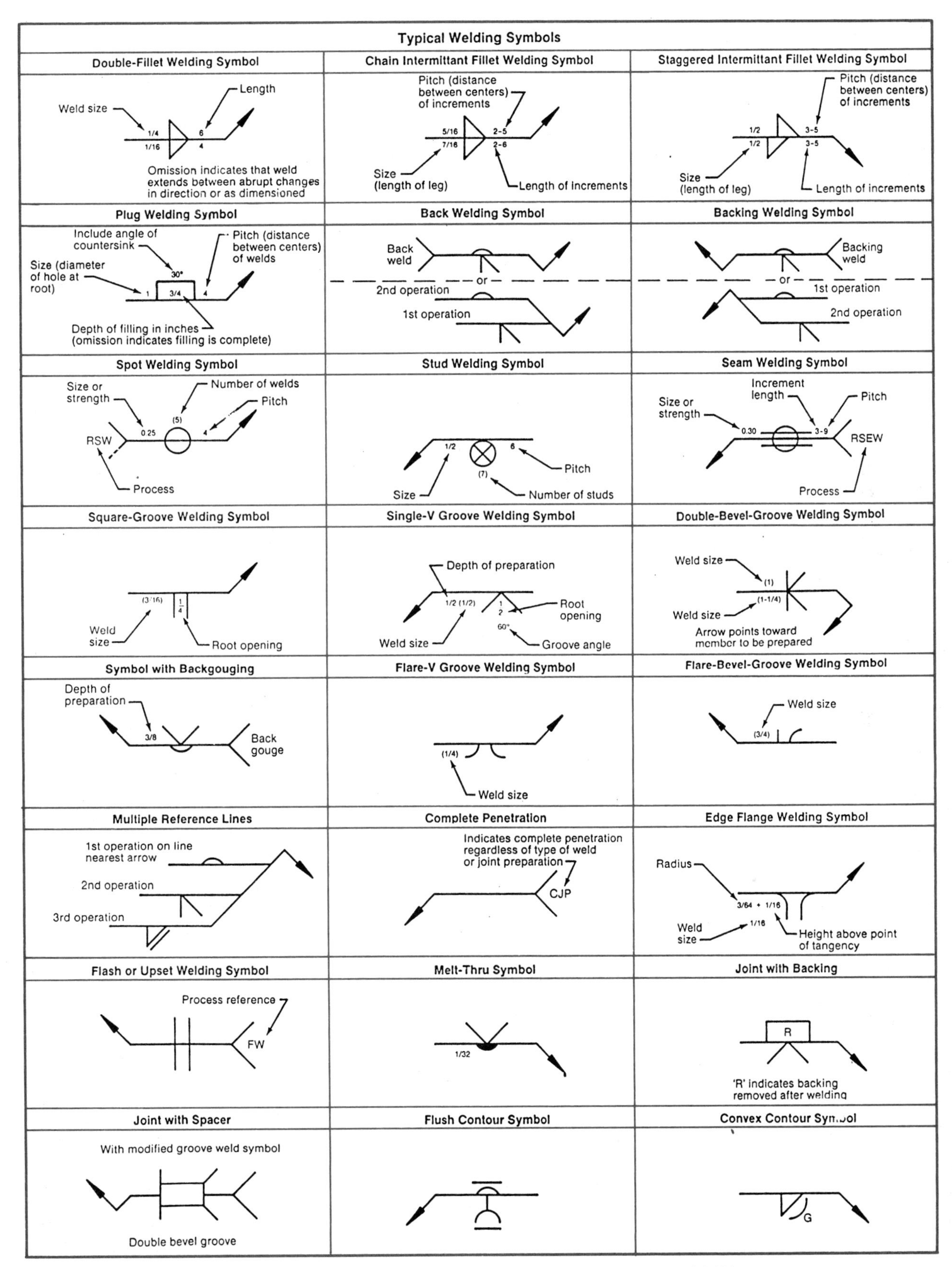

*It should be understood that these charts are intended only as shop aids. The only complete and official presentation of the standard welding symbols is in A2.4.

Fig. 15-3.

CHEMICAL TREATMENTS FOR REMOVAL OF OXIDE FILMS FROM ALUMINUM SURFACES				
SOLUTION	**CONCENTRATION**		**PROCEDURE**	**PURPOSE**
Nitric Acid	50% water, 50% nitric acid, technical grade.		Immersion 15 min. Rinse in cold water, then in hot water. Dry.	Removal of thin oxide film for fusion welding.
Sodium hydroxide **(caustic soda)** **followed by**	5% Sodium hydrox- ide in water.		Immersion 10-60 seconds. Rinse in cold water.	Removal of thick oxide film for all welding processes.
Nitric acid	Concentrated		Immerse for 30 seconds. Rinse in cold water, then hot water. Dry.	Removal of thick oxide film for all welding processes.
Sulfuric- **chromic**	H_2SO_4 CrO_3 Water	1 gal. 45 oz. 9 gal.	Dip for 2-3 min. Rinse in cold water, then hot water. Dry.	Removal of films and stains from heat treating, and oxide coatings.
Phosphoric- **chromic**	H_3PO_3 (75%) CrO_3 Water	3.5 gal. 1.75 lbs. 10 gals.	Dip for 5-10 min. Rinse in cold water. Rinse in hot water. Dry.	Removal of anodic coatings.

Fig. 15-4. Oxide films, if allowed to remain on aluminum surfaces, will affect the quality of the weld. Refer also to Chapter 10 for additional instructions.

INCHES PER POUND OF WIRE													
Wire Diameter													
Decimal	**Fraction**	**Mag.**	**Alum.**	**Alum. Bronze (10)%**	**Stain-less Steel**	**Mild Steel**	**Stain-less Steel**	**Si. Bronze**	**Copper Nickel**	**Nickel**	**De-ox. Copper**	**Ti**
.020		50500	32400	11600	11350	11100	10950	10300	9950	9900	9800	19524
.025		34700	22300	7960	7820	7680	7550	7100	6850	6820	6750	12492
.030		22400	14420	5150	5050	4960	4880	4600	4430	4400	4360	8776
.035		16500	10600	3780	3720	3650	3590	3380	3260	3240	3200	6372
.040		12600	8120	2900	2840	2790	2750	2580	2490	2480	2450	4884
.045	3/64	9990	6410	2290	2240	2210	2210	2040	1970	1960	1940	3852
.062	1/16	5270	3382	1220	1180	1160	1140	1070	1040	1030	1020	2028
.078	5/64	3300	2120	756	742	730	718	675	650	647	640	
.093	3/32	2350	1510	538	528	519	510	480	462	460	455	964.8
.125	1/8	1280	825	295	289	284	279	263	253	252	249	499.92

Fig. 15-5.

WELD METAL REQUIREMENTS FOR FILLET WELDS				
Size of Fillet	45 Deg. Fillets		30-60 Deg. Fillets	
Inches	Lbs. of Metal Per Foot	Lbs. of Rod Per Foot	Lbs. of Metal Per Foot	Lbs. of Rod Per Foot
1/8	.027	.039	.054	.078
3/16	.063	.090	.126	.180
1/4	.106	.151	.212	.302
5/16	.166	.237	.332	.474
3/8	.239	.342	.478	.684
7/16	.325	.465	.650	.930
1/2	.425	.607	.850	1.214
5/8	.663	.948	1.226	1.896
3/4	.955	1.364	1.910	2.728
7/8	1.300	1.857	2.600	3.714
1	1.698	2.425	3.396	4.850

Fig. 15-6.

TABLE 1

CURRENT RANGES FOR GMAW OF STEEL WITH SHORT-ARC TRANSFER				
	WELDING CURRENT			
	FLAT POSITION		VERTICAL AND OVERHEAD POSITIONS	
WIRE DIAMETER	MINIMUM	MAXIMUM	MINIMUM	MAXIMUM
.030 inches	50	150	50	125
.035 inches	75	175	75	150
.045 inches	100	225	100	175

Fig. 15-7. Current ranges for welding steel with short-arc transfer depend on wire diameter and the welding position.

TABLE 2

SPRAY-ARC TRANSITION CURRENTS FOR ELECTRODE TYPES			
ELECTRODE TYPE	DIA.	SHIELDING GAS	MINIMUM CURRENT (AMPERES)
MILD STEEL	.030	98% Ar-2% Oxygen	150
MILD STEEL	.035	98% Ar-2% Oxygen	165
MILD STEEL	.045	98% Ar-2% Oxygen	220
MILD STEEL	.062	98% Ar-2% Oxygen	275
STAINLESS STEEL	.035	99% Ar-1% Oxygen	170
STAINLESS STEEL	.045	99% Ar-1% Oxygen	225
STAINLESS STEEL	.062	99% Ar-1% Oxygen	285
ALUMINUM	.030	ARGON	95
ALUMINUM	.045	ARGON	135
ALUMINUM	.062	ARGON	180
DEOXIDIZED COPPER	.035	ARGON	180
DEOXIDIZED COPPER	.045	ARGON	210
DEOXIDIZED COPPER	.062	ARGON	310
SILICON BRONZE	.035	ARGON	165
SILICON BRONZE	.045	ARGON	205
SILICON BRONZE	.062	ARGON	270

Fig. 15-8. In spray-arc transfer, current requirements depend on the electrode type, wire diameter, and shielding gas.

TABLE 3

TYPICAL ARC VOLTAGES FOR SHORT-ARC TRANSFER WITH .035 INCH DIAMETER WIRE			
METAL	**ARGON**	**75% ARGON/ 25% CARBON DIOXIDE**	**CARBON DIOXIDE**
ALUMINUM	19	—	—
MAGNESIUM	16	—	—
CARBON STEEL	—	19	20
LOW ALLOY STEEL	—	19	20
STAINLESS STEEL	—	21	—
COPPER	24	—	—
COPPER NICKEL	23	—	—
SILICON BRONZE	23	—	—
ALUMINUM BRONZE	23	—	—
PHOSPHOR BRONZE	23	—	—

Fig. 15-9.

TABLE 4

MILD STEEL AND LOW ALLOY STEEL WIRES CHEMICAL COMPOSITIONS							
AWS CLASS	**CARBON**	**MANG.**	**SILICON**	**SULPHUR**	**PHOS.**	**MOLYB.**	**OTHER**
E70S-1	.07-.19	.90-1.40	.30-.50	.035	.025	—	—
E70S-2	.06	.90-1.40	.40-.70	.035	.025	—	.05-.15 TI. .02-.12 ZR. .05-.15 AL.
E70S-3	.06-.15	.90-1.40	.45-.70	.035	.025	—	—
E70S-4	.07-.15	.90-1.40	.65-.85	.035	.025	—	—
E70S-5	.07-.19	.90-1.40	.30-.60	.035	.025	—	.50-.90 AL.
E70S-6	.07-.15	1.40-.1.85	.80-1.15	.035	.025	—	—
E70S-1B	.07-.12	1.60-.2.10	.50-.80	.035	.025	.40/.60	—

Fig. 15-10.

TABLE 5

GMAW PROCESS PARAMETERS				
STEEL MATERIAL .030 IN. DIAMETER WIRE SHORT-ARC MODE				
THICKNESS	**GAS**	**AMPERES**	**WIRE SPEED**	**VOLTS**
22 Ga.	75 Ar-25 Co_2	40-55	90-100	15-16
20 Ga.	75 Ar-25 Co_2	50-60	120-135	15-16
18 Ga.	75 Ar-25 Co_2	70-80	150-175	16-17
16 Ga.	75 Ar-25 Co_2	90-110	220-250	17-18
14 Ga.	75 Ar-25 Co_2	120-130	250-340	17-18

Fig. 15-11.

TABLE 6

GMAW PROCESS PARAMETERS				
STEEL MATERIAL	.035 IN. DIAMETER WIRE		SHORT-ART MODE	
THICKNESS	**GAS**	**AMPERES**	**WIRE SPEED**	**VOLTS**
1/8 in.	75 Ar-25 Co_2	140-150	280-300	18-19
3/16 in.	75 Ar-25 Co_2	160-170	320-340	18-19
1/4 in.	75 Ar-25 Co_2	180-190	360-380	21-22
5/16 in.	75 Ar-25 Co_2	200-210	400-420	21-22
3/8 in.	75 Ar-25 Co_2	220-250	420-520	23-24

Fig. 15-12.

TABLE 7

GMAW PROCESS PARAMETERS				
STEEL MATERIAL	.030 IN. DIAMETER WIRE		SHORT-ARC MODE	
THICKNESS	**GAS**	**AMPERES**	**WIRE SPEED**	**VOLTS**
22 Ga.	Co_2	40-55	90-100	16-17
20 Ga.	Co_2	50-60	120-135	17-18
18 Ga.	Co_2	70-80	150-175	18-19
16 Ga.	Co_2	90-110	220-250	19-20
14 Ga.	Co_2	120-130	250-340	20-21

Fig. 15-13.

TABLE 8

GMAW PROCESS PARAMETERS				
STEEL MATERIAL	.035 IN. DIAMETER WIRE		SHORT-ARC MODE	
THICKNESS	**GAS**	**AMPERES**	**WIRE SPEED**	**VOLTS**
1/8 in.	Co_2	140-150	280-300	21-22
3/16 in.	Co_2	160-170	320-340	21-22
1/4 in.	Co_2	180-190	360-380	23-24
5/16 in.	Co_2	200-210	400-420	23-24
3/8 in.	Co_2	220-250	420-520	24-25

Fig. 15-14.

TABLE 9

GMAW PROCESS PARAMETERS				
STEEL MATERIAL	.045 IN. DIAMETER WIRE		SHORT-ARC MODE	
THICKNESS	**GAS**	**AMPERES**	**WIRE SPEED**	**VOLTS**
1/8 in.	75 Ar-25 Co_2	140-150	140-150	18-19
3/16 in.	75 Ar-25 Co_2	160-170	160-175	18-19
1/4 in.	75 Ar-25 Co_2	180-190	185-195	21-22
5/16 in.	75 Ar-25 Co_2	200-210	210-220	21-22
3/8 in.	75 Ar-25 Co_2	220-250	220-270	23-24

Fig. 15-15.

TABLE 10

GMAW PROCESS PARAMETERS				
STEEL MATERIAL	.035 IN. DIAMETER WIRE		SPRAY-ARC MODE	
THICKNESS	**GAS**	**AMPERES**	**WIRE SPEED**	**VOLTS**
1/8 in.	98 Ar-2 Oxygen	160-170	320-340	23-24
3/16 in.	98 Ar-2 Oxygen	180-190	360-380	24-25
1/4 in.	98 Ar-2 Oxygen	200-210	400-420	24-25
5/16 in.	98 Ar-2 Oxygen	220-250	420-520	25-26

Fig. 15-16.

TABLE 11

GMAW PROCESS PARAMETERS				
STEEL MATERIAL	.045 IN. DIAMETER WIRE		SPRAY-ARC MODE	
THICKNESS	**GAS**	**AMPERES**	**WIRE SPEED**	**VOLTS**
1/8 in.	98 Ar-2 Oxygen	160-170	160-175	23-24
3/16 in.	98 Ar-2 Oxygen	180-190	185-195	24-25
1/4 in.	98 Ar-2 Oxygen	200-210	210-220	24-25
5/16 in.	98 Ar-2 Oxygen	220-250	220-270	25-26
3/8 in.	98 Ar-2 Oxygen	300 up	375 up	26-27
1/2 in.	98 Ar-2 Oxygen	315 up	390 up	29-30

Fig. 15-17.

TABLE 12

GMAW PROCESS PARAMETERS				
STAINLESS STEEL MATERIAL	.035 IN. DIAMETER WIRE		SHORT-ARC MODE	
THICKNESS	**GAS**	**AMPERES**	**WIRE SPEED**	**VOLTS**
18 Ga.	Tri-Mix	50-60	120-150	19-20
16 Ga.	Tri-Mix	70-80	180-205	19-20
14 Ga.	Tri-Mix	90-110	230-275	20-21
12 Ga.	Tri-Mix	120-130	300-325	20-21
3/16 In.	Tri-Mix	140-150	350-375	21-22
1/4 in.	Tri-Mix	160-170	400-425	21-22
5/16 in.	Tri-Mix	180-190	450-475	21-22

Fig. 15-18.

TABLE 13

GMAW PROCESS PARAMETERS				
STAINLESS STEEL MATERIAL	.035 IN. DIAMETER WIRE		SPRAY-ARC MODE	
THICKNESS	**GAS**	**AMPERES**	**WIRE SPEED**	**VOLTS**
3/16 in.	98 Ar-2 Oxygen	160-170	400-425	23-24
1/4 in.	98 Ar-2 Oxygen	180-190	450-475	24-25

Fig. 15-19.

TABLE 14

GMAW PROCESS PARAMETERS				
STAINLESS STEEL MATERIAL	1/16 IN. DIAMETER WIRE		SPRAY-ARC MODE	
THICKNESS	**GAS**	**AMPERES**	**WIRE SPEED**	**VOLTS**
3/8 in.	98 Ar-2 Oxygen	220-250	As Req'd.	25-26
7/16 in.	98 Ar-2 Oxygen	300 up	As Req'd.	26-27
1/2 in.	98 Ar-2 Oxygen	325 Up	As Req'd.	27-32

Fig. 15-20.

TABLE 15

GMAW PROCESS PARAMETERS				
ALUMINUM MATERIAL .035 IN. DIAMETER WIRE SPRAY-ARC MODE				
THICKNESS	GAS	AMPERES	WIRE SPEED	VOLTS
1/8 in.	Argon	110-130	350-400	21-22
3/16 in.	Argon	140-150	425-450	23-24

Fig. 15-21.

TABLE 16

GMAW PROCESS PARAMETERS				
ALUMINUM MATERIAL 3/64 IN. DIAMETER WIRE SPRAY-ARC MODE				
THICKNESS	GAS	AMPERES	WIRE SPEED	VOLTS
3/16 in	Argon	140-150	300-325	23-24
1/4 in.	Argon	180-210	350-375	24-25
5/16 in.	Argon	200-230	400-425	26-27
3/8 in.	Argon	220-250	450-480	26-28

Fig. 15-22.

TABLE 17

GMAW PROCESS PARAMETERS				
ALUMINUM MATERIAL 1/16 IN. DIAMETER WIRE SPRAY-ARC MODE				
THICKNESS	GAS	AMPERES	WIRE SPEED	VOLTS
1/4 in.	Argon	180-210	170-185	24-25
5/16 in.	Argon	200-230	200-210	26-27
3/8 in.	Argon	220-250	220-230	26-28
7/16 in.	Argon	280 up	240-270	28-29
1/2 in.	Argon	300 up	290-300	29-30

Fig. 15-23.

ETCHING REAGENTS FOR MICROSCOPIC EXAMINATION OF IRON AND STEEL			
APPLICATION	**ETCHING**	**COMPOSITION**	**REMARKS**
Carbon, Low-Alloy and Medium-Alloy Steels	1. Nital	Nitric acid (sp gr 1.42). 1-5 ml Ethyl or Methyl alcohol. 95-99 ml	Darkens pearlite, and gives contrast between adjacent colonies; reveals ferrite boundaries; differentiates ferrite from martensite; shows case depth of nitrided steel. Etching time: 5-60 secs
	2. Picral	Picric acid . . . . 4 g Methyl alcohol . . . . 100 ml	Used for annealed and quench-hardened carbon and alloy steel. Not as effective as No. 1 for revealing ferrite grain boundaries. Etching time: 5-120 secs
	3. Hydrochloric and picral acids	Hydrochloric acid . . . . . 5 ml Picric acid . . . . 1 g Methyl alcohol . . . . 100 ml	Reveals austenitic grain size in both quenched and quenched-and-tempered steels
Alloy and Stainless Steels	4. Mixed acids	Nitric acid . . . . 10 ml Hydrochloric acid . . . . 20 ml Glycerol . . . . 20 ml Hydrogen Peroxide . . . . 10 ml	Iron-Chromium-Nickel-and Manganese alloy steel. Etching: use fresh acid
	5. Ferric Chloride Chloride	Ferric Chloride . . . . 5 g Hydrochloric acid . . . . . 20 ml Water, distilled . . . . 100 ml	Reveals structure of stainless and austenitic nickel steels
	6. Marble's Reagent	Cupric Sulfate . . . . 4 g Hydrochloric acid . . . . . 20 ml Water, distilled . . . . 20 ml	Reveals structure of various stainless steels
High Speed Steels	7. Snyder-Graff	Hydrochloric acid . . . . . 9 ml Nitric acid . 9 ml Methyl alcohol . . . . 100 ml	Reveals grain size of quenched and tempered high speed steels. Etching time: 15 secs to 5 min

Fig. 15-24

ETCHING PROCEDURES			
REAGENTS	**COMPOSITION**	**PROCEDURE**	**USES**
Solutions For Aluminum			
Sodium Hydroxide	NaOH 1 gm H_2O 99 ml	Swab 10 seconds	General microscopic
Tucker's Etch	HF 15 ml HCl 45 ml HNO_3 15 ml H_2O 25 ml	Etch by immersion	Macroscopic
Solutions For Stainless Steel			
Nitric & Acetic Acids	HNO_3 30 ml Acetic Acid 20 ml	Apply by swabbing	For stainless alloys and others high in nickel or cobalt.
Cupric sulphate	$CuSO_4$ 4 gms HCl 20 ml H_2O 20 ml	Etch by immersion	Structure of stainless steels
Cupric chloride & Hydrochloric acid	$CuCl_2$ 5 gms HCl 100 ml Ethyl Alcohol 100 ml H_2O 100 ml	Use cold immersion or swabbing	For austenitic & ferritic steels

Fig. 15-25

ETCHING PROCEDURES			
REAGENTS	**COMPOSITION**	**PROCEDURE**	**USES**
Solutions For Copper And Brass			
Ammonium Hydroxide & Ammonium Persulphate	NH_4OH 1 part H_2O 1 part $(NH_4)_2S_2O_3$ (2 1/2%) 2 parts	Immersion	Polish attack of copper and some alloys.
Chromic Acid	Saturated aqueous solution (CrO_3)	Immersion or swabbing	Copper, brass, bronze, nickel silver (plain etch)
Ferric Chloride	$FeCl_3$ 5 parts HCl 10 parts H_2O 100 parts	Immersion or swabbing (etch lightly)	Copper, brass, bronze, aluminum bronze.
Solutions For Iron & Steel			
Macro Examination			
Nitric Acid	HNO_3 5 ml H_2O 95 ml	Immerse 30 to 60 seconds	Shows structure of welds.
Ammonium persulphate	$(NH_4)_2S_2O_3$ 10 gms H_2O 90 ml	Surface should be rubbed with cotton during etching	Brings out grain structure, recry-stallization at welds.
Nital	HNO_3 5 ml Ethyl Alcohol 95 ml	Etch 5 min. followed by 1 sec. in HCl (10%)	Shows cleanness Depth of hardening carburized or decarburized sur-faces, etc.
Micro Examination			
Picric acid (Picral)	Picric acid 4 gms Ethyl or methyl alcohol (95%) 100 ml	Etching time a few seconds to a minute or more	For all grades of carbon steels.

Fig. 15-26

TABLE OF ALLOY WIRE WEIGHTS AND MEASURES (Approx.)						
Brown & Sharpe Gage No.	Decimal	Phos. Bronze Ft. per lb.	18% Nickel Ft. per lb.	Aluminum Ft. per lb.	Copper Ft. per lb.	Brass Ft. per lb.
4/0	.4600	1.559	1.599	5.207	1.561	1.640
3/0	.4096	1.966	2.016	6.567	1.968	2.068
2/0	.3648	2.480	2.543	8.279	2.482	2.068
1/0	.3249	3.127	3.206	10.44	3.130	2.608
1	.2893	3.943	4.043	13.16	3.947	3.289
2	.2576	4.972	5.098	16.60	4.977	4.147
3	.2294	6.269	6.428	20.94	6.276	5.229
4	.2043	7.906	8.106	26.40	7.914	6.594
5	.1819	9.969	10.22	22.20	9.980	8.315
6	.1620	12.57	12.89	41.98	12.58	10.49
7	.1443	15.85	16.25	52.91	15.87	13.22
8	.1285	19.99	20.49	66.73	20.01	16.67
9	.1144	25.20	25.84	84.19	25.23	21.02
10	.1019.	31.78	32.59	106.1	31.82	26.51
11	.09074	40.08	41.09	133.9	40.12	33.43
12	.08081	50.53	51.82	168.8	50.59	42.15
13	.07196	63.72	65.39	212.5	63.80	53.15
14	.06408	80.35	82.34	268.2	80.44	66.88
15	.05707	101.3	103.9	337.9	101.4	84.68
16	.05082	127.8	131.0	426.9	127.9	106.6
17	.04526	161.1	165.2	536.9	161.3	134.4
18	.04030	203.2	208.3	678.4	203.4	169.7
19	.03589	256.2	262.7	854.9	256.5	213.7
20	.03196	323.	331.2	1076.	323.4	269.5
21	.02846	407.3	417.7	1356.	407.8	339.8
22	.02535	513.6	526.7	1721.	514.2	428.5
23	.02257	647.7	664.1	2157.	648.4	540.2
24	.02010	816.7	837.4	2727.	817.7	681.3
25	.01790	1030.	1056.	3439.	1031.	859.
26	.01594	1299.	1332.	4358.	1300.	1083.
27	.0142	1638.	1679.	5464.	1639.	1366.
28	.01264	2065.	2117.	6940.	2067.	1723.

Fig. 15-27

HARDNESS CONVERSION TABLE

BRINELL			ROCKWELL			
Dia. in mm, 3000 kg. load 10 mm ball	Hardness No.	Vickers or Firth Hardness No.	C 150 kg. load 120° Diamond Cone	B 100 kg. load 1/16 in. dia. ball	Scleroscope No.	Tensile Strength 1000 psi
2.05	898					440
2.10	857					420
2.15	817					401
2.20	780	1150	70		106	384
2.25	745	1050	68		100	368
2.30	712	960	66		95	352
2.35	682	885	64		91	337
2.40	653	820	62		87	324
2.45	627	765	60		84	311
2.50	601	717	58		81	298
2.55	578	675	57		78	287
2.60	555	633	55	120	75	276
2.65	534	598	53	119	72	266
2.70	514	567	52	119	70	256
2.75	495	540	50	117	67	247
2.80	477	515	49	117	65	238
2.85	461	494	47	116	63	229
2.90	444	472	46	115	61	220
2.95	429	454	45	115	59	212
3.00	415	437	44	114	57	204
3.05	401	420	42	113	55	196
3.10	388	404	41	112	54	189
3.15	375	389	40	112	52	182
3.20	363	375	38	110	51	176
3.25	352	363	37	110	49	170
3.30	341	350	36	109	48	165
3.35	331	339	35	109	46	160
3.40	321	327	34	108	45	155
3.45	311	316	33	108	44	150
3.50	302	305	32	107	43	146
3.55	293	296	31	106	42	142
3.60	285	287	30	105	40	138
3.65	277	279	29	104	39	134
3.70	269	270	28	104	38	131
3.75	262	263	26	103	37	128
3.80	255	256	25	102	37	125
3.85	248	248	24	102	36	122
3.90	241	241	23	100	35	119
3.95	235	235	22	99	34	116
4.00	229	229	21	98	33	113
4.05	223	223	20	97	32	110
4.10	217	217	18	96	31	107
4.15	212	212	17	96	31	104
4.20	207	207	16	95	30	101
4.25	202	202	15	94	30	99
4.30	197	197	13	93	29	97
4.35	192	192	12	92	28	95
4.40	187	187	10	91	28	93
4.45	183	183	9	90	27	91
4.50	179	179	8	89	27	89
4.55	174	174	7	88	26	87
4.60	170	170	6	87	26	85
4.65	166	166	4	86	25	83
4.70	163	163	3	85	25	82
4.75	159	159	2	84	24	80
4.80	156	156	1	83	24	78
4.85	153	153		82	23	76
4.90	149	149		81	23	75
4.95	146	146		80	22	74
5.00	143	143		79	22	72
5.05	140	140		78	21	71
5.10	137	137		77	21	70
5.15	134	134		76	21	68
5.20	131	131		74	20	66
5.25	128	128		73	20	65
5.30	126	126		72		64
5.35	124	124		71		63
5.40	121	121		70		62

Continued page 129

HARDNESS CONVERSION TABLE						
BRINELL			ROCKWELL			
Dia. in mm, 3000 kg. load 10 mm ball	Hardness No.	Vickers or Firth Hardness No.	C 150 kg. load 120° Diamond Cone	B 100 kg. load 1/16 in. dia. ball	Scleroscope No.	Tensile Strength 1000 psi
5.45	118	118		69		61
5.50	116	116		68		60
5.55	114	114		67		59
5.60	112	112		66		58
5.65	109	109		65		56
5.70	107	107		64		56
5.75	105	105		62		54
5.80	103	103		61		53
5.85	101	101		60		52
5.90	99	99		59		51
5.95	97	97		57		50
6.00	95	95		56		49

Fig. 15-28 continued.

PROPERTIES OF ELEMENTS AND METAL COMPOSITIONS						
Elements	Symbol	Density (specific gravity)	Weight per cu. ft.	Specific heat	Melting point °C.	Melting point °F.
Aluminum	Al	2.7	166.7	0.212	658.7	1217.7
Antimony	Sb	6.69	418.3	0.049	630	1166
Armco iron	...	7.9	490.0	0.115	1535	2795
Carbon	C	2.34	219.1	0.113	3600	6512
Chromium	Cr	6.92	431.9	0.104	1615	3034
Columbium	Cb	7.06	452.54	...	1700	3124
Copper	Cu	8.89	555.6	0.092	1083	1981.4
Gold	Au	19.33	1205.0	0.032	1063	1946
Hydrogen	H	0.070*	0.00533	...	−259	−434.2
Iridium	Ir	22.42	1400.0	0.032	2300	4172
Iron	Fe	7.865	490.9	0.115	1530	2786
Lead	Pb	11.37	708.5	0.030	327	621
Manganese	Mn	7.4	463.2	0.111	1260	2300
Mercury	Hg	13.55	848.84	0.033	−38.7	−37.6
Nickel	Ni	8.80	555.6	0.109	1452	2645.6
Nitrogen	N	0.97*	0.063	...	−210	−346
Oxygen	O	1.10*	0.0866	...	−218	−360
Phosphorus	P	1.83	146.1	0.19	44	111.2
Platinum	Pt	21.45	1336.0	0.032	1755	3191
Potassium	K	0.87	54.3	0.170	62.3	144.1
Silicon	Si	2.49	131.1	0.175	1420	2588
Silver	Ag	10.5	655.5	0.055	960.5	1761
Sodium	Na	0.971	60.6	0.253	97.5	207.5
Sulfur	S	1.95	128.0	0.173	119.2	246
Tin	Sn	7.30	455.7	0.054	231.9	449.5
Titanium	Ti	5.3	218.5	0.110	1795	3263
Tungsten	W	17.5	1186.0	0.034	3000	5432
Uranium	U	18.7	1167.0	0.028		
Vanadium	V	6.0	343.3	0.115	1720	3128
Zinc	Zn	7.19	443.2	0.093	419	786.2
Bronze (90 percent Cu 10 percent Sn)	...	8.78	548.0	...	850–1000	1562–1832
Brass (90 percent Cu 10 percent Zn)	...	8.60	540.0	...	1020–1030	1868–1886
Brass (70 percent Cu 30 percent Zn)	...	8.44	527.0	...	900–940	1652–1724
Cast pig iron	...	7.1	443.2	...	1100–1250	2012–2282
Open-hearth steel	...	7.8	486.9	...	1350–1530	2462–2786
Wrought-iron bars	...	7.8	486.9	...	1530	2786

*Density compared with air.

Linde Division, Union Carbide Corp.

Fig. 15-29

WEIGHTS AND EXPANSION PROPERTIES				
Metal	Weight per ft³ (lbs.)	Weight per m³ (kg)	Expansion per °F rise in temperature (.0001 in.)	Expansion per °C rise in temperature (.0001 mm)
Aluminum	165	2643	1.360	62.18
Brass	520	8330	1.052	48.10
Bronze	555	8890	.986	45.08
Copper	555	8890	.887	40.55
Gold	1200	19222	.786	35.94
Iron (Cast)	460	7369	.556	25.42
Lead	710	11373	1.571	71.83
Nickel	550	8810	.695	31.78
Platinum	1350	21625	.479	21.90
Silver	655	10492	1.079	49.33
Steel	490	7849	.689	31.50

Fig. 15-30

SAFE LIMITS FOR WELDING FUMES			
MATERIAL	GASES		MILLION PARTS per Cu. Ft.
	ppm	mg/m³	
Acetylene	1000		
Beryllium		.002	
Cadmium Oxide fumes		.1	
Carbon Dioxide	5000		
Copper fumes		.1	
Iron Oxide fumes		10.0	
Lead		.2	
Manganese		5.0	
Nitrogen Dioxide	5.0		
Oil Mist		5.0	
Ozone	.1		
Titanium Oxide		15.0	
Zinc Oxide fumes		5.0	
Silica, crystalline			2.5
Silica, amorphous			20.0
Silicates:			
Asbestos			5.0
Portland Cement			50.0
Graphite			15.0
Nuisance Dust			50.0

Fig. 15-31

GENERAL METRIC/U.S. CONVENTIONAL CONVERSIONS			
Property	**To convert from**	**To**	**Multiply by**
acceleration (angular)	revolution per minute squared	rad/s²	1.745 329 × 10^{-3}
acceleration (linear)	in./min²	m/s²	7.055 556 × 10^{-6}
	ft/min²	m/s²	8.466 667 × 10^{-5}
	in./min²	mm/s²	7.055 556 × 10^{-3}
	ft/min²	mm/s²	8.466 667 × 10^{-2}
	ft/s²	m/s²	3.048 000 × 10^{-1}
angle, plane	deg	rad	1.745 329 × 10^{-2}
	minute	rad	2.908 882 × 10^{-4}
	second	rad	4.848 137 × 10^{-6}
area	in.²	m²	6.451 600 × 10^{-4}
	ft²	m²	9.290 304 × 10^{-2}
	yd²	m²	8.361 274 × 10^{-1}
	in.²	mm²	6.451 600 × 10^{2}
	ft²	mm²	9.290 304 × 10^{4}
	acre (U.S. Survey)	m²	4.046 873 × 10^{3}
density	pound mass per cubic inch	kg/m³	2.767 990 × 10^{4}
	pound mass per cubic foot	kg/m³	1.601 846 × 10
energy, work, heat, and impact energy	foot pound force	J	1.355 818
	foot poundal	J	4.214 011 × 10^{-2}
	Btu*	J	1.054 350 × 10^{3}
	calorie*	J	4.184 000
	watt hour	J	3.600 000 × 10^{3}
force	kilogram-force	N	9.806 650
	pound-force	N	4.448 222
impact strength	(see energy)		
length	in.	m	2.540 000 × 10^{-2}
	ft	m	3.048 000 × 10^{-1}
	yd	m	9.144 000 × 10^{-1}
	rod (U.S. Survey)	m	5.029 210
	mile (U.S. Survey)	km	1.609 347
mass	pound mass (avdp)	kg	4.535 924 × 10^{-1}
	metric ton	kg	1.000 000 × 10^{3}
	ton (short, 2000 lbm)	kg	9.071 847 × 10^{2}
	slug	kg	1.459 390 × 10
power	horsepower (550 ft lbf/s)	W	7.456 999 × 10^{2}
	horsepower (electric)	W	7.460 000 × 10^{2}
	Btu/min*	W	1.757 250 × 10
	calorie per minute*	W	6.973 333 × 10^{-2}
	foot pound-force per minute	W	2.259 697 × 10^{-2}
pressure	pound force per square inch	kPa	6.894 757
	bar	kPa	1.000 000 × 10^{2}
	atmosphere	kPa	1.013 250 × 10^{2}
	kip/in.²	kPa	6.894 757 × 10^{3}
temperature	degree Celsius, $t°_C$	K	$t_K = t°_C + 273.15$
	degree Fahrenheit, $t°_F$	K	$t_K = (t°_F + 459.67)/1.8$
	degree Rankine, $t°_R$		$t_K = t°_R/1.8$
	degree Fahrenheit, t_F		$t°_C = (t_F - 32)/1.8$
	kelvin, t_K		$t°_C = t_K - 273.15$
tensile strength (stress)	ksi	MPa	6.894 757
torque	inch pound force	N·m	1.129 848 × 10^{-1}
	foot pound force	N·m	1.355 818

Fig. 15-32

GENERAL METRIC/U.S. CONVENTIONAL CONVERSIONS (continued)			
Property	**To convert from**	**To**	**Multiply by**
velocity (angular)	revolution per minute	rad/s	$1.047\ 198 \times 10^{-1}$
	degree per minute	rad/s	$2.908\ 882 \times 10^{-4}$
	revolution per minute	deg/min	$3.600\ 000 \times 10^{2}$
velocity (linear)	in./min	m/s	$4.233\ 333 \times 10^{-4}$
	ft/min	m/s	$5.080\ 000 \times 10^{-3}$
	in./min	mm/s	$4.233\ 333 \times 10^{-1}$
	ft/min	mm/s	5.080 000
	mile/hour	km/h	1.609 344
volume	in.^{3}	m^3	$1.638\ 706 \times 10^{-5}$
	ft^3	m^3	$2.831\ 685 \times 10^{-2}$
	yd^3	m^3	$7.645\ 549 \times 10^{-1}$
	in.^{3}	mm^3	$1.638\ 706 \times 10^{4}$
	ft^3	mm^3	$2.831\ 685 \times 10^{7}$
	in.^{3}	L	$1.638\ 706 \times 10^{-2}$
	ft^3	L	$2.831\ 685 \times 10$
	gallon	L	3.785 412

*thermochemical

Fig. 15-32 continued.

METRIC UNITS FOR WELDING		
Property	**Unit**	**Symbol**
area dimensions	square millimeter	mm^2
current density	ampere per square millimeter	A/mm^2
deposition rate	kilogram per hour	kg/h
electrical resistivity	ohm meter	$\Omega \cdot$ m
electrode force (upset, squeeze, hold)	newton	N
flow rate (gas and liquid)	liter per minute	L/min
fracture toughness	meganewton meter$^{-3/2}$	MN·m$^{-3/2}$
impact strength	joule	J = N·m
linear dimensions	millimeter	mm
power density	watt per square meter	W/m^2
pressure (gas and liquid)	kilopascal	kPa = 1000 N/m^2
tensile strength	megapascal	MPa = 1000 000 N/m^2
thermal conductivity	watt per meter kelvin	W/(m·K)
travel speed	millimeter per second	mm/s
volume dimensions	cubic millimeter	mm^3
wire feed rate	millimeter per second	mm/s

Fig. 15-33

CONVERTING MEASUREMENTS FOR COMMON WELDING PROPERTIES			
Property	**To convert from**	**To**	**Multiply by**
area dimensions (mm²)	in²	mm²	$6.451\ 600 \times 10^{2}$
	mm²	in²	$1.550\ 003 \times 10^{-3}$
current density (A/mm²)	A/in²	A/mm²	$1.550\ 003 \times 10^{-3}$
	A/mm²	A/in²	$6.451\ 600 \times 10^{2}$
deposition rate** (kg/h)	lb/h	kg/h	.045**
	kg/h	lb/h	2.2**
electrical resistivity (Ω · m)	Ω · cm	Ω · m	$1.000\ 000 \times 10^{-2}$
	Ω · m	Ω · cm	$1.000\ 000 \times 10^{2}$
electrode force (N)	pound-force	N	4.448 222
	kilogram-force	N	9.806 650
	N	lbf	$2.248\ 089 \times 10^{-1}$
flow rate (L/min)	ft³/h	L/min	$4.719\ 475 \times 10^{-1}$
	gallon per hour	L/min	$6.309\ 020 \times 10^{-2}$
	gallon per minute	L/min	3.785 412
	cm³/min	L/min	$1.000\ 000 \times 10^{-3}$
	L/min	ft³/h	2.118 880
	cm³/min	ft³/h	$2.118\ 880 \times 10^{-3}$
fracture toughness ($MN \cdot m^{-3/2}$)	$ksi \cdot in.^{1/2}$	$MN \cdot m^{-3/2}$	1.098 855
	$MN \cdot m^{-3/2}$	$ksi \cdot in.^{1/2}$	0.910 038
heat input (J/m)	J/in.	J/m	$3.937\ 008 \times 10$
	J/m	J/in.	$2.540\ 000 \times 10^{-2}$
impact energy	foot pound force	J	1.355 818
linear measurements (mm)	in.	mm	$2.540\ 000 \times 10$
	ft	mm	$3.048\ 000 \times 10^{2}$
	mm	in.	$3.937\ 008 \times 10^{-2}$
	mm	ft	$3.280\ 840 \times 10^{-3}$
power density (W/m²)	W/in.²	W/m²	$1.550\ 003 \times 10^{3}$
	W/m²	W/in.²	$6.451\ 600 \times 10^{-4}$
pressure (gas and liquid) (kPa)	psi	Pa	$6.894\ 757 \times 10^{3}$
	lb/ft²	Pa	$4.788\ 026 \times 10$
	N/mm²	Pa	$1.000\ 000 \times 10^{6}$
	kPa	psi	$1.450\ 377 \times 10^{-1}$
	kPa	lb/ft²	$2.088\ 543 \times 10$
	kPa	N/mm²	$1.000\ 000 \times 10^{-3}$
	torr (mm Hg at 0 °C)	kPa	$1.333\ 22 \times 10^{-1}$
	micron (μm Hg at 0 °C)	kPa	$1.333\ 22 \times 10^{-4}$
	kPa	torr	$7.500\ 64 \times 10$
	kPa	micron	$7.500\ 64 \times 10^{3}$
tensile strength (MPa)	psi	kPa	6.894 757
	lb/ft²	kPa	$4.788\ 026 \times 10^{-2}$
	N/mm²	MPa	1.000 000
	MPa	psi	$1.450\ 377 \times 10^{2}$
	MPa	lb/ft²	$2.088\ 543 \times 10^{4}$
	MPa	N/mm²	1.000 000
thermal conductivity (W/[m·K])	cal/(cm·s·°c)	W/(m·K)	$4.184\ 000 \times 10^{2}$
travel speed, wire feed speed (mm/s)	in./min	mm/s	$4.233\ 333 \times 10^{-1}$
	mm/s	in./min	2.362 205

*Preferred units are given in parentheses.
**Approximate conversion.

Fig. 15-34

DICTIONARY OF WELDING TERMS

A

ACETONE: Colorless, volatile, water-soluble, flammable liquid used to remove grease and oils from weld joint prior to welding.

ALLOY: A material formed by the combination of two or more metallic elements.

ALTERNATING CURRENT: Electrical current is the flow of electrons through a conductor. In alternating current, the electrons flow in one direction, stop, and reverse the flow (alternate directions) for each complete cycle. In the United States, alternating current flows at 60 cycles per second. Therefore, the current stops and starts 120 times a second.

AMERICAN IRON AND STEEL INSTITUTE (AISI): Industry association of iron and steel producers. It provides statistics on steel production and use. Publishes steel products manuals.

AMERICAN WELDING SOCIETY (AWS): Nonprofit, technical society organized and founded for the purpose of advancing the art and science of welding. The AWS publishes codes and standards concerning all phases of welidng plus a magazine, *The Welding Journal.*

AMMETER: Instrument for measuring electrical current in amperes.

AMPERAGE: Strength of an electrical current measured in amperes.

ANNEAL: Removal of internal stresses in metal by heating and slow cooling.

ARC BLOW: Deflection of intended arc pattern by magnetic fields.

ASPHYXIATION: Loss of consciousness due to lack of oxygen.

AUSTENITE: Nonmagnetic stainless steel that cannot be hardened by heat treatment. This type of steel is characterized by its unique grain structure. Contains at least 11 percent chromium with varying amounts of nickel.

AUTOMATIC WELDING: Welding that is completely controlled by a series of controllers without the aid of a welding operator.

B

BACKUP BAR: Tool or fixture attached to the root of weld joint. Tool may or may not control the shape of the penetrating metal.

BERYLLIUM: Hard, light metallic element used in copper alloys for better fatigue endurance.

BEVEL: Angular type of edge preparation.

BORE: Inside diameter of hole, tube, or hollow object.

BRIGHT METAL: Material preparation where the surface has been ground or machined to a bright surface to remove scale or oxides.

BURNTHROUGH: Weld which has melted through, resulting in a hole and excessive penetration.

BUTTERING: Form of surfacing where one or more layers of metal are placed on the weld groove face. Material then becomes transition weld deposit for final weld joint.

C

CADMIUM: White ductile metallic element used for plating material to prevent corrosion.

CARBIDES: Compound of carbon with one or more metal elements.

CARBIDE PRECIPITATION: A condition occurring in stainless steels from overheating where chromium and carbon in the grains migrate to the grain boundaries. This makes the steel more vulnerable to corrosion and weld failure.

CERTIFIED CHEMICAL ANALYSIS: Report of chemical analysis on particular heat, lot, or section of material.

CHEMICAL COMPOSITION: Composition of a material in chemical percentages.

CLADDING: Layer of material applied to a surface for the purpose of improved corrosion resistance.
COLD LAPS: Area of weld that has not fused with the base material.
CONCAVE WELD CROWN: Weld crown that is curved inward.
CONTAMINATION: Indicates a dirty part, impure shielding gas, or impure filler material.
CONTOUR: Shape of the weld bead or pass.
CONVEX WELD CROWN: Weld crown that is curved outward.
CORRECTIVE ACTION: Action to be taken to prevent weld discrepancies.
CORROSION: Eating away of material by a corrosive medium.
CRACKING: Operation of blowing dirt out of a high pressure gas cylinder valve.
CRATER: Depression at the end of a weld that has insufficient cross section.
CRATER CRACKS: Cracking that occurs in the crater.

D

DEMURRAGE: Monetary charge applied to the user of gas cylinders beyond the agreed rental period.
DEOXIDIZED FILLER MATERIALS: Filler materials which contain deoxidizers such as aluminum, zirconium, and titanium for welding steels.
DEPARTMENT OF TRANSPORTATION (DOT): Government organization responsible for establishing and maintaining rules and safety precautions for the safe handling of fuels and gases used in welding.
DESTRUCTIVE TESTING (DT): Series of tests by destruction to determine the quality of a weld.
DEWARS: Specially constructed tank similar to a vacuum bottle for the storage of liquefied gases.
DIRECT CURRENT: Flow of current (electrons) in only one direction, either to the workpiece or to the electrode.
DIRECT CURRENT ELECTRODE NEGATIVE (DCEN): Direct current flowing from electrode to the work.
DIRECT CURRENT ELECTRODE POSITIVE (DCEP): Direct current flowing from work to the electrode.
DIRECT CURRENT REVERSE POLARITY (DCRP): See preferred terminology, Direct Current Electrode Positive.
DIRECT CURRENT STRAIGHT POLARITY (DCSP): See preferred terminology, Direct Current Electrode Negative.
DRIVE ROLLERS: Specially designed rollers for various types and sizes of filler wire to be fed through a mechanized wire feeder.
DROSS: Oxidized metal or impurities within the metal.
DUCTILITY: Property of material causing it to deform permanently, or to exhibit plasticity without breaking while under tension.

E

ELECTRONIC POTENTIOMETERS: Electrical devices using electronic circuits to control and regulate electrical current.
ELONGATION: Permanent elastic extension which metal undergoes during tensile testing. Amount of extension is usually indicated by percentages of original gauge length.

F

FERROUS METALS: Group of metals containing substantial amounts of iron.
FILLET WELD: Weld of approximately triangular cross section joining two surfaces approximately at right angles in a lap joint, T joint, or corner joint.
FILLET WELD LEG: Leg length of largest isosceles right triangle which can be inscribed within fillet weld cross section.
FLOW METER: Mechanical device used for measuring inert gas rate of flow. Usually measurement is in cubic feet per hour (CFH).

H

HARD SURFACING: Hard material applied to surface of softer material for protection from abrasion or wear.

I

INERT GAS: Gas which does not normally combine chemically with the base metal or filler material.
INTERPASS TEMPERATURE: In multiple pass weld, minimum or maximum temperature specified for the deposited metal before the next weld pass is started.

M

MAGNETIC ARC BLOW: See arc blow.
MARTENSITE: Structure obtained when steel is heated and cooled to achieve its maximum hardness.

O

OUT-OF-POSITION WELDING: Welding that is performed in a nonstandard way such as vertical or overhead.

OXIDE FILM: Film formed on base material as a result of exposure to oxidizing agents, atmosphere, chemicals, or heat.

P

POLARITY: Direction of current. Current moving from the electrode to the workpiece is DCEN or DCSP. Current flow from the workpiece to the electrode is DCEP or DCRP.

POROSITY: Pores within a weld caused by gas entrapment during solidification of weld metal.

POSTHEAT: Heat which is applied at the end of the weld cycle to slow down cooling rate to prevent cracking and to relieve stress.

S

SPATTER: Small pieces of metal which have been ejected from molten pool and attached to base material outside the weld.

SPOT WELD: Controlled weld cycle to produce sheet metal weld with specific characteristics.

STRINGER BEAD: Weld bead made without oscillation (side-to-side motion).

SURFACING: Applying material to the surface of another material for protection from chemicals, heat, wear, rust, etc.

T

TACKWELD: Weld made to hold parts of a weldment in alignment until final weld is made.

TENSILE TEST: A destructive test where a weld is pulled apart. This test determines how much tension a weld can withstand before the weld gives.

TOLERANCE: Permissible variation of a characteristic, variable, or parameter.

W

WASH BEAD: Weld beads made with an oscillation (side-to-side) technique to widen the weld bead.

WHISKERS: Pieces of weld wire which have penetrated through the weld joint and melted. The wire extends beyond the penetration on the root side of the weld.

WROUGHT MATERIAL: Material made by processes other than casting.

INDEX

Q

R

S